The Merely Personal

The Merely Personal

OBSERVATIONS ON SCIENCE AND SCIENTISTS

Jeremy Bernstein

IVAN R. DEE
Chicago 2001

 For information, address: Ivan R. Dee, Publisher, 1332 North Halsted Street, Chicago 60622. Manufactured in the United States of America and printed on acid-free paper.

Library of Congress Cataloging-in-Publication Data:
Bernstein, Jeremy, 1929–
The merely personal : observations on science and scientists / Jeremy Bernstein.
p. cm.
Includes bibliographical references and index.
ISBN 1-56663-344-3 (alk. paper)
1. Science. 2. Scientists. I. Title.
Q171 .B5375 2001
500—dc21 00-063917

Acknowledgments

Some of the essays in this collection have appeared elsewhere in greater or lesser altered forms. In the former category are "The Bead in the Glass House," a portion of which was published in the *American Scholar,* and "Enough Einstein?" which appeared with a somewhat different selection of books in the *American Journal of Physics.* Less altered are "Heaven's Net," "The Merely Very Good," and "The Six Pieces of Richard Feynman," all of which appeared in the *American Scholar,* "Kurt Gödel," "The German Atomic Bomb," and "Nash," all of which appeared in *Commentary,* and "Shadows," which appeared in the *New York Review of Books.* The other essays have never been published. I would like to thank Joseph Epstein, my editor of many years at the *American Scholar,* for all the encouragement he gave me, and Ivan Dee, for once again taking on a writing project with skill and sensitivity.

Contents

"It is quite clear to me that the religious paradise of youth, which was thus lost [by the reading of popular science books], was a first attempt to free myself from the chains of the 'merely personal,' from an existence which is dominated by wishes, hopes, and primitive feelings. Out yonder there was this huge world, which exists independently of us human beings and which stands before us like a great, eternal riddle. . . ."

—Albert Einstein

"But what I admired most about Michele [Besso] was the fact that he was able to live so many years with one woman, not only in peace but in constant unity, something I have lamentably failed at twice."

—Albert Einstein

The Merely Personal

ON SCIENCE

As I was reading over the essays that I wanted to include in this collection, I made a curious discovery. I have been writing about science for the general public for some four decades, but, as far as I can remember, I have never written a "popular science" article. By a popular science article I mean an article that attempts to explain some scientific subject in its own terms, without reference to the people and circumstances that produced it. All my articles on science involve people and places. I have been interested in this human side of science ever since I got involved with it. No one ever said to me that my articles should have people and places in them. It was just the way I thought about things. That is probably why my way of writing about these matters fitted so well with the *New Yorker*—at least until the balance shifted from science to people and places. Thus I do not see a very clear division between the articles I include in this section and the next except that in this section the weight is more on the science and in the next it is more on the people and places. But it is really a continuum.

The first article in the section is divided into two parts. The first deals with the chess match played in Iceland in the summer of 1972 between Bobby Fischer and Boris Spassky. As the reader will discover, it was a scene that borders on the indescribable. The second part deals with the chess match played between the current world champion, Gary Kasparov, and a machine—IBM's Deep Blue. Kasparov lost. Perhaps we all lost, for reasons I will explain. I will also explain my own interest in these chess-playing machines and how they work, which goes back a long way. I have owned several, including some of the early ones in which moves were signaled on an actual board when a light flashed on one of the squares. The present generation of "machines" are really computer programs. They don't make obvious mistakes, and if you make one they will, as the chess players like to say, cut you up like a chicken. That is what happened to poor Kasparov. I notice he has not been eager for a rematch.

The second entry in this section grew out of an invitation I received in the fall of 1994 to write a short commentary for the *New Theater Review,* a publication of the Lincoln Center Theater in New York. The theater was putting on a play by Tom Stoppard called *Hapgood.* Remarkably, this play had been inspired by Stoppard's reading about quantum mechanics. While I am second to no one in my admiration for Stoppard, having read the play I felt the connection was, to put it mildly, overdone. My dilemma was that I had only about a thousand words to explain this. I did the best I could, but the result, which was published in their fall issue, didn't really satisfy me. I then sat down and wrote what I really wanted to say without worrying about the word length. That is what I have included here. It has never been published before.

The final two selections in this section deal with, respectively, the cosmic and the cosmologic. The cosmic concerns the explosion that was witnessed in February 1987 of an unprepossessing blue star known as Sanduleak -69° 202. In a relatively short time it became a supernova (a very large and luminous explosion), which was then known as 1987A. I was especially interested in this because of what happened the day before the astronomers actually "saw" the explosion—the day before the visible light, which had been traveling for some 170,000 years, got here. The light was preceded by the arrival of a cohort of ghostly neutrinos—"ghostly" because they scarcely interact with anything. That is why they escaped the detritus of the exploding star before the light, which took some time to diffuse through the matter. That they were detected was a bit of an accident. Experiments were under way to detect neutrinos but for an entirely different purpose. I describe both the experiments and the purpose in the essay. It was these experiments that rather accidentally detected the cosmic neutrinos. What I do not describe in the essay is the neutrino itself. I will fill in this lacuna here.

The neutrino—although not the name—was invented in 1931 by the Austrian-born theoretical physicist Wolfgang Pauli. Pauli was responding to an apparent paradox in radioactive decays in which electrons are emitted. These are what physicists call beta decays. The prototypical beta decay is one in which a neutron decays into a slightly less massive proton and an electron. One can measure the various energies of the electrons produced in this decay when an entire sample of neutrons is studied. It turns out that this distribution of energies has a very characteristic shape. Most significantly, there is not a single energy but a continuum. This was very puzzling because one observed only two particles—the electron and proton—after the decay. There was much consternation over this conundrum. Niels Bohr even made the radical—and incorrect—suggestion that energy might not be conserved. Pauli then solved the problem, but at a price. The price was the suggestion that an unobserved, electrically neutral third particle was emitted in the decay, and that it carried off the missing energy. Pauli was so embarrassed by this idea that he never published it. He presented it in a letter which was then circulated. One of the readers was Enrico Fermi, who created the first neutrino theory of beta decay and named the particle "the little neutral one" in Italian. For the next twenty-odd years the neutrino had a very peculiar role in elementary particle physics. Everyone agreed that it must exist, though no one had seen one. Thus it was always included in theories of beta decay but with a certain sense of embarrassment. That all changed in 1955–1956.

At this time two experimental physicists, Frederick Reines and Clyde Cowan, were able to make use of a very large nuclear reactor at the Savannah River Plant in South Carolina. The spent fuel elements from this reactor contain radioactive isotopes that beta decay. These decays produce a huge flux of neutrinos—actually anti-neutrinos. I will discuss the difference shortly. Some 10^{13} anti-neutrinos per square centimeter per second are emitted. For our purposes, what is important is that when an anti-neutrino collides with a proton it can produce a neutron and a positive electron—the positron. The positron wanders around until it finds a stray electron that it annihilates, producing a very characteristic burst of radiation. This is what Cowan and Reines observed, but not often. Because of the weakness of the neutrino's interactions, they saw only some three events an hour. It took many months to collect enough events to be sure that there was an effect. Now, with large accelerators, it is routine to produce beams of neutrinos and anti-neutrinos. In any event, after 1956 there was no question that the neutrino existed. Then came the Glorious Revolution.

The Glorious Revolution began in the summer of 1956 and was in full cry by early 1957. It involved the overthrow of what had been an accepted canon of symmetry in physics—the symmetry of left and right, also known as parity symmetry. This is a bit abstract, but in essence what it said was that you could make a coordinate system—a frame of reference—using three fingers of your right hand or three fingers of your left hand held at right angles, and it didn't matter in terms of the physical laws. But it turned out that it *did* matter. The weak interactions that produced beta decay were not left-right symmetric. Much of the theoretical work done on this was done by two Chinese-American physicists, T. D. Lee and C. N. Yang. My first serious scientific profile for the *New Yorker* was a dual profile of them and an explanation of the Glorious Revolution. I think it was the first really serious scientific article the magazine published. The new results opened up some dramatic possibilities. It turned out that if the neutrino masses were exactly zero, one could revive a very elegant theory of the neutrino that the mathematician Hermann Weyl had created many years earlier but which had been abandoned since it seemed to violate parity symmetry. This theory suggested that as a neutrino moved, it spun in a fixed direction while the anti-neutrino spun in the opposite direction. (This is a bit homespun, but I don't want to go into the question of the spin of elementary particles here.)

All of this was very unexpected and therefore very exciting, but there was more to come. In 1962 it was shown that there was more than one kind of neutrino—more than one "flavor," as one would now say. The neutrinos emitted with ordinary electrons are called for obvious reasons electron neutrinos. But there is a heavy electron that is known as the muon. Neutrinos emitted with it are called muon neutrinos, and in 1962 it was shown that they are a distinct kind of particle. Thirteen years later it was shown that there was a third kind of neutrino which physicists call the tau neutrino since it is emitted with a still heavier electron, the tau. As far as we know, this is the extent of it. It is possible that we have found all the types of neutrinos that exist. But the fact that there were these distinct types raised a new possibility. The Weyl theory required that the neutrino have exactly zero mass. Experiments showed that at least the electron neutrino had a tiny mass, if it had any at all. But no one could find any good theoretical reason why the mass should be exactly zero. When the new neutrinos were discovered there seemed to be even less reason. Hence people began exploring the consequences of the neutrinos having small masses. One of these consequences is very remarkable. It turns out that if you produce a neutrino of a given flavor—say, an electron neutrino—then if, say, the muon neutrino has a different mass, some of the electron neutrinos can convert themselves into muon neutrinos or tau neutrinos. This is a testable proposition because you can look for the different flavors as the neutrinos evolve in time. Indeed, this has been done and is still being done. It appears as if this transformation actually does take place. This would imply that at least one of the neutrinos has a mass. Pauli, what have you wrought! Now to the cosmology.

In 1917 Einstein produced the first modern cosmological theory. As I discuss in the essay, his view of what the universe was is totally different from our own and totally different from the universe as Einstein came to know it four decades later. Einstein's 1917 universe consisted of only our Milky Way galaxy, and this, he decided, was stationary, neither expanding nor contracting. But he found that he could not produce such a universe with his new theory of gravitation, which suggested that all massive objects attract each other gravitationally. Thus he modified his theory by introducing a new force of unknown origin which counteracted gravity in the large and kept the universe stationary. This he called the "cosmologic member," and its strength is characterized by a number we call the cosmological constant. In my essay I trace the rise and fall of this constant and its possible resurrection. Einstein came to believe that its introduction was his biggest scientific "blunder." But was it? Read the essay and decide.

The Bead Game in the Glass House

I. 1972

"I imagine," Knecht wrote to his patron, "that one can be an excellent Glass Bead player, even a virtuoso, and perhaps even a thoroughly competent Magister Ludi, without having any inkling of the real mystery of the Game and its ultimate meaning. It might even be that one who does guess or know the truth might prove a greater danger to the Game, were he to become a specialist in the Game, or a Game leader. For the dark interior, the esoterics of the Game, points down into the One and All, into those depths where the eternal Atman eternally breathes in and out, sufficient unto itself. One who had experienced the ultimate meaning of the Game within himself would by that fact no longer be a player; he would no longer dwell in the world of multiplicity and would no longer be able to delight in invention, construction and combination, since he would know altogether different joys and raptures. Because I think I have come close to the meaning of the Glass Bead Game, it will be better for me and for others if I do not make the Game my profession, but instead shift to music."—Hermann Hesse, *The Glass Bead Game*

In the fall of 1971 I won a senior National Science Foundation Fellowship. These fellowships—which I think no longer exist—allowed university professors in the sciences to take a year off from teaching and to do research any place in the world that seemed suitable. I had always wanted to study at either Oxford or Cambridge, and as I knew some of the senior physicists at Oxford that is where I chose to go. But, as it happened, that spring there was a coal miner's strike in England. We had no heat. It got so cold in my office that I worked with my gloves on. One morning, as I was sitting there trying to work in my mittened fingers, the phone rang. Much to my surprise it was a call from Chicago. I was still more surprised when the amiable voice on the other end announced that he was an editor of *Playboy* magazine. (No, it was not Himself.) He asked if in fact I was the Jeremy Bernstein who had written the profile of Stanley Kubrick for the *New Yorker.* Having been assured that I was the very Bernstein, he made a remarkable proposition. I was to go to Iceland, he said, all expenses paid, for as long as I needed, to gather material to write an article on the world championship chess match that was to take place that summer in Reykjavik between Boris Spassky and Bobby Fischer. He also told me that the magazine had hired, as its expert, the American grand master Larry Evans. I would have at my disposal for the entire match the services of my very own grand master.

After hanging up the phone I had two reactions: the first was an enormous attraction to the idea of going to Iceland, and the second was an equally enormous repulsion at the idea of having my name appear in *Playboy.* By 1972 I had been writing for the *New Yorker* for over a decade, and what had begun as a "hobby" had turned into a serious avocation. I had literally two professions, and many of my physicist colleagues were beginning to think of me as something of a farceur. What, then, would they think if I appeared in *Playboy*? Gone would be my fellowships and contracts with the National Science Foundation. I finally resolved this dilemma by proposing to *Playboy* that I write the chess piece under an assumed name. In fact I had one already picked out: Jay Amber, "amber" being the English translation of the German "bernstein." (Sometime later I learned that Leonard Bernstein had written some of his very earliest compositions under the pseudonym L. Amber.) I had hit upon this name when I first began writing for the *New Yorker.* I thought, even then, that writing for a popular magazine which had—erroneously—the reputation of being simply "funny" might cause me problems. I explained this to the *New Yorker*'s editor, William Shawn. He said I could use any name I liked, but that if I wrote under an assumed name, in his view, I would come to regret it. I dropped the idea. No one ever found the works of "Jay Amber" in *Playboy* either. For reasons I will explain, my article was never published.

In any event, the first game of the Fischer-Spassky match was scheduled to begin in Reykjavik on the 2nd of July. As it turned out, on that day Fischer was not in Iceland but in Queens, New York, holding out for more money. I was neither in Iceland nor in Queens, nor in fact in Oxford. I was in Borehamwood chez Kubrick, where on Sunday nights he used to show films in his private projection room. We had taken time out from our film viewing to watch a BBC documentary on Bobby Fischer which was called "This Little Thing with Me and Spassky." It was a rather ordinary documentary, modestly enlivened by some shots showing Fischer crushing a twelve-year-old chess prodigy named Lewis Cohen. He also refused to eat a birthday cake which had been offered to him by Mike Wallace. "I don't eat this kind of cake," Fischer commented. Kubrick and I were sort of nodding off when we were mesmerized by the following childhood tale: Fischer as a young child was taught to play chess by his sister, Joan Fischer Targ. She was six years older. When Fischer was six, she had given him his first chess set. Fischer was raised largely by his mother and sister after his father had deserted the family. Very rapidly Bobby began to best his sister, so they took to exchanging sides when Fischer got ahead. But there were times when his sister was not around, so Fischer played against himself—playing both black and white. As he was describing this he added softly, but very distinctly, and with no trace of humor, "Mostly I won."

Not long after this interview, a London investment banker named Jim Slater, who, though he never set foot in Iceland during the match, had declared himself to be a devotee of the game, put up an estimated fifty thousand pounds toward the prize for the match. The total purse, the largest ever offered in chess up to that point, was $125,000, to be split sixty-forty between winner and loser. According to championship rules, twenty-four games were to be played, with the winner accumulating the most points (one point for a game won, one-half point for a game tied).

Once Slater announced his donation, a horde descended on Reykjavik. The cast of characters included, besides the Fischer-Spassky entourages with their chess-master seconds, such literary heavy hitters as Arthur Koestler and George Steiner, who was covering the match for the *New Yorker.* In addition there were some 250 "lesser" journalists from all over the world. The *New York Times* was represented by its Pulitzer Prize–winning senior music critic Harold Schonberg, an old friend and a very enthusiastic and knowledgeable chess player. In the course of preparing a magazine article on Fischer, Schonberg told me, he played a game with him. It took nineteen moves before his position became totally hopeless. Afterward, Fischer told him that until the fourteenth move the game had been identical to one he had played some years earlier in South America. "Your fourteenth move loses in all variations," Fischer explained, which he then proceeded to illustrate with six or seven variations.

Once it became clear to me that there was a real possibility of an actual match—keep in mind that at that time Fischer had already played five games in previous matches with Spassky, the reigning world champion, and had won none of them, so that part of his antics, one thought, might have been a ploy to avoid a humiliating defeat—I headed for Iceland. I stayed, on and off, for some weeks in a delightful bed and breakfast in Reykjavik where the water, which had a slightly sulfuric smell, was heated "volcanically." Iceland is on a continental divide, and something is always popping up from the depths, including a new island that appeared that summer.

Mr. Slater's largesse had gotten Fischer as far as Reykjavik, but it soon became clear that it might not be sufficient to get Fischer to an actual chessboard. (These, incidentally, had been carved out of marble by one Sigurdur Helgason, whose normal occupation was carving tombstones. By the time Fischer actually showed up in Reykjavik, Helgason was on his fourth design, Fischer having rejected, from long distance, the first three.) After Fischer arrived in Reykjavik, on the 4th of July, he promptly went to sleep. In fact he slept through the opening ceremonies where he and Spassky were supposed to draw for color—black and white—for their first game. Fischer had already missed the first scheduled game, which should have been played the preceding Sunday. Since Spassky had shown up, and Fischer had no valid medical excuse, the Russians were quite reasonably insisting that Fischer forfeit that game. Moreover Spassky, whose good nature and tact during all of this bordered on the superhuman, considering the provocations, was now insisting that Fischer present him with a formal written apology for his behavior—something that most of us thought was as likely to happen as the levitation of Iceland into outer space. But it happened. Fischer had somehow been persuaded that it was his patriotic duty to play since the match involved the Russians. Fischer's hatred of the Russians bordered on the manic. Nonetheless a flowery "Dear Boris" note—with a stinger at the end asking, in the name of "sportsmanship," that the Russians withdraw their request for a first game forfeiture, which they didn't —was delivered to Spassky at his hotel the following morning.

With all the general kefuffle, the first game did not take place until Tuesday, July 11. It was scheduled to begin at 5 p.m. and was scheduled to last until 10, an hour that could be predicted—and which was more or less still in broad daylight, considering the latitude of Reykjavik—since each player, by the rules, was required to make forty moves in no more than two and a half hours. The Reykjavik Exhibition Hall was filled to its capacity of 2,500 people, most of them Icelanders, each of whom had paid five dollars to get in. These figures impressed me because the average annual income in Iceland at the time was estimated to be $2,500, and many people were working two jobs to make ends meet. Five dollars, therefore, was not a trivial sum of money.

First to appear on stage was Boris Spassky, suavely handsome and impeccably dressed. He was greeted by warm applause. By this time Fischer's misbehavior—which, by the way, had only just begun—was already getting on the nerves of the remarkably tolerant Icelanders. Spassky had white, and promptly at five—though Fischer had not yet appeared—he made his first move: pawn to queen four (I am partial to this now obsolescent way of designating chess moves. The more "hip" notation would be simply "d4"—progress), and started Fischer's clock. In tournament chess each player starts the other's clock after he or she makes a move. No Fischer. Perhaps he had changed his mind again and was simply not going to play. Some eight minutes went by with Fischer's clock running while Spassky walked up and down the stage. From time to time he peered at Fischer's empty chair, which, by the way, was a special leather swiveling affair that had been designed by Charles Eames and had been shipped to Reykjavik from Michigan, where it was made to order for Fischer. Spassky's own chair was an ordinary, nonswiveling, yellow office chair. (Later in the match Spassky got his own swiveling chair, and during some of the games the two of them sat swiveling at each other.) Then Fischer lumbered in. He was large, powerful-looking, and somewhat ungainly. When I first saw him I was reminded of Einstein's not very flattering characterization of the mathematician John von Neumann, whose politics may have been to the right even of Fischer's. Einstein referred to von Neumann as "ein Denktier"—a think animal.

It was clear from the beginning that Fischer's mind was not on this game. A film producer named Chester Fox thought he had secured the rights to film the match, and apparently had some sort of contract to this effect. This legal technicality did not seem to faze Fischer, who refused to have himself filmed by Chester Fox. In fact, Fox seemed to have bought these rights from a third party, with no Fischer signature in evidence. Whether money was the issue, or whether Fischer simply did not want himself filmed by anyone, is unclear. It also must be said that the camera arrangement was obtrusive. The cameras were located in towers in the auditorium itself. I thought they made a whirring noise when they were operating, and there were also people who moved in and out of the towers. It was probably the first time that a chess match like this had ever been filmed—the interest was tremendous—and not enough thought had gone into how to separate the filmers from their subjects.

In any event, the presence of the cameras—over his objections—seemed to drive Fischer over the edge. It affected his play in this game. On the twenty-ninth move he made a pawn capture with his bishop, which caused many people in the audience to wonder if he had taken leave of his senses altogether. Spassky sat bolt upright in his chair. At first he was not sure what to make of it, but a few moves later he took Fischer's bishop. Play was soon adjourned for the night—what was left of it—and the players closeted themselves with their seconds. When play resumed the next day—Fischer had allowed some thirty minutes to run off his clock—it was clear that he was going to lose. After the fifty-sixth move he resigned with a surprisingly cheerful wave at the audience. It was the most humanly appealing thing I saw him do the whole summer. He was overheard to say to someone, "Today I played like a fish, but wait until tomorrow."

There was almost no tomorrow. It was said that after the game Fischer ripped the phone wire out of the wall in his hotel room so that he wouldn't have to speak to anyone. It was also rumored that afterward he had gone to the American air base in Keflavik, some thirty miles away, and spent the night bowling. There appeared to be an absolute impasse. Fischer would not play if Chester Fox filmed him, and Chester Fox claimed that he had a contract. He was willing to remove the towers, but the cameras would stay. Nonetheless the next game had been scheduled for five o'clock the following day. The hall was once again full. This time Fischer had white and the first move. But there was no Fischer. As the rules specified, promptly at five o'clock the referee—a German named Lothar Schmidt—started Fischer's clock. According to the rules, Fischer had one hour to make his first move, after which time the game was forfeited. We all sat there for an hour, including the Icelanders who had paid five dollars apiece to watch this. At the end of the hour we all marched out of the hall in silence. Fischer had lost, and no one had seen anything. It was remarkable that the Icelanders simply did not tear the place down.

At this point it would have been very difficult to find anyone who thought that the match would continue. Fischer was now down zero games to two. (Somehow the first game, which should have been played while Fisher was in Queens, had been written off.) And in the next game he would have black—a decided disadvantage because white establishes the attack. Fischer was insisting that Fox's cameras be removed, and to back this up he had produced a baroque legal document—which of course he had never signed—that he claimed gave him an absolute veto over any filming arrangement. He now had six hours, according to the rules, from the end of the game—or nongame—which had terminated at 6 p.m., to make a formal complaint or risk a total forfeit. Promptly at midnight a lengthy document handwritten by

Fischer was delivered to Lothar Schmidt outlining Fischer's grievances. He called Fox's camera people "bungling unknowns who claimed to be professional cameramen" and noted that they were "clumsy, rude and deceitful." He ended by saying that if all the camera equipment were removed from the hall, "I will be at the chessboard." A committee was formed, consisting of various representatives from the Fischer-Spassky camps and the Icelanders, to decide what to do. The next day a press conference was called, which we all attended.

It was a remarkable occasion. The Russians spoke in Russian and were interpreted. Fischer was represented by one Fred Cramer, a retired lighting engineer from Milwaukee who claimed that the mere existence of the cameras bothered Fischer, however expertly they were manipulated. What, Cramer argued, would happen if Fischer wanted to scratch himself or "pick his nose"? The Russians asked how many more solitary matches Spassky would have to endure before being declared the champion and collecting the winner's share of the purse. Three, it was decided. It turned out that if Fischer refused to play, his contract specified that he would get none of the loser's share. The committee had decided to try to convince Fischer to play the next game, which was to take place a couple of days later, by meeting with him face to face.

This raised a new set of problems. Although Fischer, like Spassky and indeed like Gary Kasparov, is half Jewish, it had been reported that he had recently joined a fundamentalist Christian sect called the Worldwide Church of God which was based in Pasadena, California. Fischer had paid a visit to Ambassador College, which was maintained by the sect in Pasadena, but had left before being baptized—a requisite for membership. Nonetheless he had adopted the sect's view of the Sabbath. From sundown on Friday to sundown on Saturday he was, it was alleged, shut off from the world with his Bible. The committee had, then, until sundown Friday to see Fischer. But when, indeed, was sundown in July in Iceland? Having witnessed it, I can assure you that it is nearly broad daylight there until close to midnight. In this matter Fischer had been uncharacteristically helpful and had decided to define "sundown" as occurring at 11:45 p.m. So at 11:45 we were all gathered in the Loftleider Hotel—Fischer's quarters—to hear Lothar Schmidt's very pessimistic evaluation of his discussion with Fischer. Meanwhile a Scandinavian journalist with a knowledge of Norwegian and hence Icelandic, a form of old Norse, had heard a participant on an Icelandic talk show suggest that Fischer and all the foreign journalists on the island should be given eight hours to get out or be shot. That was enough for me. I was on the first flight the following morning back to London.

I began this account with a quotation from Herman Hesse's *Glass Bead Game.* Here is another. "Our game is neither philosophy nor religion; it is a discipline of its own, in character most akin to art." Akin to art. Yes, great chess is akin to art. A beautiful chess game or chess problem offers, in its own way, as much aesthetic satisfaction as a great painting or a great novel. Over the next several weeks Fischer's play against Spassky was indeed akin to art. One was almost inclined to overlook everything else. After all, one does not need to remind oneself of Van Gogh's ear every time one looks at one of his paintings. But how did this happen? We last saw Fischer locked up in his hotel room, about to begin the observance of his Sabbath, having rejected the latest entreaties from Lothar Schmidt. How did we get from there to here?

In looking back at it, the key may have lain in Fischer's comportment during Lothar Schmidt's nocturnal visit at the Hotel Loftleider. Schmidt reported that Fischer was pleasant and appeared lucid but was absolutely determined to get his own way. This meant that at least he was connected. He was also being subjected to enormous pressures internally and externally. The internal pressure came from the fact that, as Schmidt had explained to him, and as he must have known himself, if he pulled out of this tournament in this way he would never be invited to play in another one. This meant that he could never realize his ambition to become the recognized world champion. Spassky would be world champion, and no amount of self-aggrandizement on Fischer's part could change that.

Externally, he received a second phone call from Henry Kissinger, apparently a chess player. The first time Kissinger called was when Fischer was still in Queens. When asked if he had called under instructions from President Nixon, Kissinger remarked, "I just called him to say that a lot of people were rooting for him," and that the president was "delighted" that he had made the call. There is no evidence that Nixon had much, if any, interest in chess, but he did have an interest in the American air base in Keflavik, which was an important part of the Arctic missile early warning system. The lease on the base was about to expire, and many Icelanders, even before the present invasion, had decided that the Americans might be better off somewhere else. But like the monk who gave eight reasons why he left the order, the eighth being "They kicked me out," the "eighth reason" for Fischer's change of heart was that the Icelandic Chess Federation gave in and ordered all the cameras removed. Fischer had won.

This gave poor Chester Fox, whom I came to like, essentially nothing to do. Nonetheless he remained in Reykjavik, to which I had returned once it became clear that there was going to be a match after all. Like most of us, Chester used to hang out at night in the lobby of the Loftleider. One night about midnight I was sitting with him when the Fischer entourage came out of the elevator, got into the Mercedes with an automatic shift that Fischer had demanded, and headed off presumably for their nocturnal bowling at the Keflavik air base. After they had gone through the revolving door Chester asked me, "Is Fischer that good?" I told him that Fischer was so good that even he did not know how good he was. Then Chester said, "He'd better not see you sitting with me if you want an interview with him." Although the last thing in the world that I wanted to do was to interview

Fischer—what could one possibly ask him?—I said, "Why is that, Chester?" "He hates me," Chester replied. Studying the somewhat bedraggled and forlorn Foxian figure, it was hard to imagine him as the focus of a really powerful Fischer "hate." "You see," Chester went on, "after the first game I happened to meet him, and he said to me, 'It's nothing personal. I would do the same thing to anyone who got in my way.' I said to him, 'Bobby, you are a very sick boy and should see a doctor.' " A light dawned.

Once it became clear that Fischer was not only going to win this match but was cutting Spassky up like a chicken, there arrived in Reykjavik a second group of hustlers who made Chester Fox look like a child with a lemonade stand. These people, some of whom I will describe shortly, had two things in common: they were out to make a fast buck, and they assumed that Fischer was too. They couldn't believe that he would act in a way that they believed would get in the way of what they thought Fischer's self-interest ought to be. This is a mistake often made in diplomacy when one is dealing with a regime that simply will not behave as it should if only it conformed to our notions of reasonableness. What usually happens in this case is that the diplomats can be found at piers, or airports, shaking their heads in bewilderment as they prepare to leave the country while the natives continue to do their thing. This is what happened here as well.

The first such operator to appear was Chet Forte, who was one of the prime movers in the development of the coverage of sports on television—for example, ABC's Game of the Week and Wide World of Sports. He turned out to be a dynamic, swarthy-looking man, given to wearing elasticized double-knit slacks with matching turtlenecks that failed to conceal a considerable paunch. From a distance—New York—he had decided that whatever the problem was in Reykjavik with Fischer and the cameras, he could fix it in a matter of hours and could then show portions of the match on Wide World of Sports. He brought with him a cameraman named Lorne Hassan, one of whose credits included filming Bing Crosby fishing for salmon in Iceland. Forte had got it into his head that if he could get Fischer interested in the details of the filming—how the cameras were going to be placed and the like—Fischer might become so intrigued that the filming could proceed. This was sort of like pacifying some unruly native by letting him play with a toy. "Bobby," Forte noted, "is immature about a lot of facts of life. He doesn't know how a TV works or anything. And he's impossible to get to, but once you sit down with him you can change your opinion of him." Indeed, after having "sat down," Forte reported that Fischer had told him, "I would love more than ever to have this great match brought out to my millions of fans, but the cameras must be placed in a way they do not disturb my game."

Forte then made an interesting arrangement with Chester Fox, who by now had been reduced to selling still photographs of the match to interested newsmen. Fox would "contract out" the filming to Hassan and ABC, which would then show the film on Wide World of Sports. Parts of Chester's film of the first game had already run on Wide World. In this way the eighth game was filmed. The only problem was that no one had, it seems, bothered to inform Fischer. When he learned about it he threw a tantrum that seemed to have caught Hassan by surprise. "I don't understand it," he said. "I saw Bobby during the night, and he gave me his word that cameras could film this great match. . . . We are now waiting for instructions." They were not long in coming, and shortly thereafter Hassan and everyone else connected with ABC left on the next plane and were never seen again.

Among the viewers of ABC's showing of the first game was Jerry Weintraub, who once described himself to me as being one of the top five "skimmers" in the world—a skimmer being someone who sets up a deal for an entertainer and then skims off a percentage of the take. At the time he and his partner, Sid Bernstein (no relation), did their skimming out of a New York company called Management III. They handled, among others, Elvis Presley, Charles Aznavour, and the Moody Blues. Weintraub had watched Wide World on his farm in Maine where he was vacationing with his wife, the singer Jane Morgan. Bernstein was in Hawaii vacationing with his family. Neither man knew a fianchetto from a fettuccine, but they did know a good thing when they saw it, and by the usual rules of the game Fischer appeared to be a very good thing indeed.

Their only entrée to Fischer was through an entertainment lawyer named Paul Marshall, who shuttled back and forth between New York and Reykjavik as Fischer's representative when things began to get out of hand. While waiting for Marshall to set up an appointment, Bernstein and Weintraub joined the rest of us in the lobby of the Loftleider. They unfolded before us their vision of the future which included a Bobby Fischer long-playing record, a television spectacular, and merchandise of all sorts, all of this culminating in a series of super challenge matches—a kind of chess Super Bowl. This seemed to me at the time to be a total chimera, the sort that had led many a gold prospector to an untimely death in the hills. But Weintraub and Bernstein were not easy to discourage. A few days later Weintraub had a discussion with Fischer in his hotel suite. They were, Weintraub told me, physically separated by a partially closed door. This did not impede the acoustics, so Weintraub was able to describe to Fischer the vast sums of money that would be his if only he let himself be led down the golden path by Bernstein and Weintraub. Remarkably, some of this actually came to pass—which is why Bernstein and Weintraub take vacations in Hawaii and have farms in Maine. The fact that Chester Fox was threatening to—and indeed did—sue Fischer for breach of contract seemed of little moment. As far as I can tell, the case never came up in court. By the time a trial date had been set in 1979, Fischer had more or less disappeared.

Sometime during that Icelandic summer Harold Schonberg, who had known Fischer ever since he was a teenager, gave me two predictions. The first was more or less obvious by mid-summer—that Fischer would win the match. He did, after twenty-one games, by a score of 12.5 to 8.5. Draws count half a point for each player. The score was as close as it was because Fischer had lost the first two games and, after he had Spassky thoroughly beaten, settled for as many quick draws as possible. The second prediction, which has also turned out to be correct, seemed to me at the time to be entirely crazy—namely, that Fischer would never play another chess match. The reason I thought this was crazy was that, at least for a while, Fischer seemed to relish the adulation. He appeared on a Bob Hope television special and was given a Bobby Fischer Day in New York. True, when he was offered a key to the city he replied, "I live here, what do I need a key for?" But Schonberg knew his man. He knew that Bobby had no place to go but down. Sometime before the match he had given an interview to a reporter in which he had said, "Americans like a winner. If you lose you are nothing." If you don't play, you can't lose, and Fischer never played another match.

II. 1997

And every year at the Ludus solemnis we can see scholars of distinction who rather looked down on us Glass Bead Game Players during their work-filled year, and who have not always wished our institution well. In the course of the Great game we see them falling more and more under the spell of our art; we see them growing eased and exalted, rejuvenated and fired, until at last, their hearts strengthened and deeply stirred, they bid goodbye with words of almost abashed gratitude.—Hermann Hesse, *The Glass Bead Game*

On the afternoon of May 11, 1997, the reigning world chess champion Gary Kasparov gave a brief interview to the press. He was, to put it mildly, not a happy man. He had just been beaten by a score of 3.5 to 2.5 in a match he played with a machine—IBM's Deep Blue. He found this quite unacceptable. He thought the machine might somehow have cheated. He noted, "I have no idea what's happening behind the curtain. Maybe it was an outstanding accomplishment by the computer. But I don't think this machine is unbeatable." And, he added, "I personally assure you that if it starts to play competitive chess, put it in a fair contest and I personally guarantee you I will tear it to pieces . . ."—cut it up like a chicken. I was not at the interview nor at the game. I was downstairs in the lobby of the Equitable Center, a skyscraper on Seventh Avenue and 51st Street in Manhattan, in which the match had taken place. I had not got around to buying a ticket to the match until it was too late, but I sort of hung out in the lobby of the building hoping someone would show up trying to sell one. From time to time someone would emerge from the match, and those of us who were in the lobby then asked how it was going. The first report was that Kasparov, who was playing black, had a strong position. But sometime later the report was that he had played terribly and had lost. I felt, I must confess, extremely sad. As crazy as Fischer was, and is, at least he was a human being, which is what made the match in Reykjavik such a vast comic spectacle. There is nothing comic about watching a robot crush a person. It looks like just another day's work.

In the end *Playboy* never published my article about the Fischer-Spassky match. The editor who had gotten me to Iceland in the first place told me that Heffner had lost interest in chess and that "my" issue was to be devoted to backgammon—the world champion of which was at the time, incidentally, a machine. None of the old Reykjavik hands seem to have been at the Kasparov–Deep Blue contest. Spassky, who had impressed us all by his graciousness, after he lost said, "There's a new champion. I'm not sad. It's a sporting event and I lost. Bobby's the new champion. Now I must take a walk." He left the Soviet Union and when I last heard was living in Paris. I do not know precisely how he earns his living, but from time to time I see his name mentioned in connection with a chess match. He never again reached the championship level. Fischer has all but disappeared. Sightings have been reported in Los Angeles, where it is said that he puts anti-Semitic literature on car windshields. After he won he was asked how it felt to be the new world champion. Ever gracious he replied, "I always felt I was the champion." Poor Chester Fox died in the fall of 1996. In the obituaries I read of him he was referred to as a "film producer." I never found any mention of the role of nonfilm producer that he played in Reykjavik. By the time that match got going in earnest, bowing to Fischer's demands, it had been moved out of the main auditorium and into a back room that was usually used for table tennis tournaments. The room was fitted with a closed-circuit television so that spectators could observe the games without having any direct physical contact with the players, and vice versa.

This was the same setup that was used in Kasparov's match with the computer. Seated across from Kasparov was one of the IBM scientists who designed the machine and who made the actual board moves on its behalf. Next to the two men were little flags in stands—a Russian flag for Kasparov and an American flag for the machine. Apparently IBM does not have its own flag. After the final match I ran into one of the reporters who was covering it. He was the one who had told me the outcome. When I expressed my melancholy he said, "Well, people did create it." Somehow that did not make me feel better.

As I mentioned in the introduction, my own interest in "machine chess" goes back a long way. When I was gathering material in the early 1960s for what became my *New Yorker* profile of the computer, "The Analytical Engine," the field of electronic computers was relatively new. It dated only to World War II. I don't recall anyone that I interviewed then predicting that within a few generations machines much more powerful than the behemoths that one could then find occupying entire rooms in universities and government laboratories would become a common household appliance. One should keep in mind that the transistor was developed only in the 1950s, while the so-called "integrated circuit"—an entire electronic unit on a single chip—

was conceived in 1958 by Jack S. Kilby of Texas Instruments. Without these inventions there would be no personal computers, the first of which was designed in 1961 by the Digital Equipment Company. Nonetheless people had already begun to investigate whether a machine could perform activities normally associated with human thought—chess being a primary example.

As far as I can tell, the first person to do this seriously was Claude Shannon. Shannon, who was born in Gaylord, Michigan, in 1916, was a remarkable man. I remember seeing him around MIT in the late 1950s. He was, as I recall, tall and gaunt, and in physical type he resembled Lyle Lovett. Although I never saw him do it, he was able to ride a unicycle and juggle simultaneously. He was also one of the most important figures in the creation of what is now commonly referred to as "the information age." In fact he more or less invented what is now known as information theory. Shannon took his degree in mathematics from MIT in 1940 and the next year joined the staff of the Bell Telephone Laboratories. He returned to MIT in 1956 and remained there until his retirement. In 1949 he showed how one could make the degree of information contained in a message quantitative and thus measure the likelihood of the information being lost or garbled—something that is of enormous importance when one is trying to design a communications circuit. About the same time he wrote the seminal paper on chess-playing machines. In some sense Deep Blue and every other chess-playing machine have their origins in Shannon's paper.

The first question that Shannon dealt with is why chess is interesting—more interesting than, say, tic-tac-toe or even checkers. In some sense, chess is a determinant game. That is, if we had an infinitely powerful computer, all we would have to do is have it follow the consequences of every possible move and every possible response in any situation to guarantee a favorable outcome. Chess games—excluding such aberrations as perpetual check, a draw situation that can go on forever if one is silly enough to keep moving the king back and forth—are finite in length. So in principle we can run out the string of moves and responses and never lose. This is why chess to such a computer might seem not to be interesting and why tic-tac-toe is definitely not interesting.

The difference is in the numbers. In chess, white opens the game. There are twenty possible opening moves. Each of the eight pawns can move one or two squares—sixteen possible pawn moves—and each of the knights has two possible moves, making twenty in all. Then black has twenty possible responses. Thus in the first "ply"—a complete inventory of white and black moves from a given position to the depth of one move—there are $20 \times 20 = 400$ conceivable opening moves. But the number of possibilities immediately increases. If, in the opening, white's king pawn, for example, moves up two squares, this presents in the next ply the possibility of queen and bishop moves as well. In fact, Shannon decided that at any given point in the game there are approximately thirty moves and thirty replies, so that on average a single ply involves $30 \times 30 = 900$ possibilities. A typical chess game might last about forty moves. This was about the average number of moves in the six games Kasparov played with Deep Blue. Thus to do the omniscient analysis discussed above we would have to go out forty plies, which involves $30^{40} \times 30^{40} = 30^{80}$ possibilities. (Remember that 10^{80} is 10 followed by eighty zeros.) Now 30^{80} is about 10^{118}. To give some meaning to this absurdly large number, let us note that Deep Blue—the fastest chess machine ever built—can analyze about 200 million possible positions a second. That means that it would take Big Blue 10^{110} seconds to run the whole chain. But the age of the universe is only about—give or take a factor of ten—10^{17} seconds old. We could have had Deep Blue running at full steam since the Big Bang and we would not have made a dent in analyzing all the possible combinations in a typical chess game. That is why chess is interesting. In checkers in a typical game there are "only" 10^{40} possibilities to analyze, which is why checkers is less interesting.

But how to make this stupendous analysis? Here Shannon offered some suggestions. Adumbrating these has been what has kept constructors of chess machines busy ever since. To see the sort of thing that is involved, let us consider an extremely simplified situation. White is to play. In our simplified situation let us assume that there are only three possible moves for white, which we can call 1, 2, and 3. After each of these moves let us assume that black has only two possible responses. So if white selects move 1, black can respond with only moves 1' and 1". The choice of course is up to black. Likewise, 2 can be followed by 2' or 2", and so on for 3. Hence the "tree" is very simple. Of course we have no clue as to which of the three moves white should choose. We must also know the consequences for white if black responds with 1' or 1", and so on. We will now specify in this model what these are. In a real game the situation is almost never as clear-cut as in this example, but we can at least get an idea of what is involved. Let us suppose that 1' and 1" lead to a draw for white. Hence if white chooses 1 as its move, a draw is guaranteed since 1' or 1" are black's only choices. Let us next suppose that 2' and 2" both lead to a win for black. Hence it would be suicidal for white to choose move 2. But move 3 is more interesting. We will suppose that 3' leads to a win for black while 3" leads to a loss. If we assume that black has any analytical capacities at all, we must assume that if white chooses 3, black will choose 3', which leads to a white loss. Thus white contemplating this chain, and, assuming that black is not an idiot or suicidal, will choose 1 and accept a draw. This sort of reasoning is called in the theory of games *minimaxing*. In an actual chess machine like Deep Blue, using much more complicated criteria, it takes place millions of times a second.

This model brings out all the features of Shannon's general analysis of how in principle we should go about trying to construct such a machine. First we must instruct the machine as to how many possibilities it should consider at each ply. In this case we have considered all the possibilities. It would certainly be useful if we could bring in criteria where in certain—hopefully most—situations we could focus only on the "relevant" possibilities. Next we have to instruct the machine how many plies to consider before making the next move. In our model example we have just considered a single ply since the outcome of the game is decided after only one move. In some situations Deep Blue looked ahead as many as eleven plies. In general, the farther we look ahead—the greater the number of plies—the better will be the machine's decisions. This of course must be balanced against the time it takes. Each player, machine or not, in tournament chess must make a specified number of moves in a specified time—typically forty moves in two hours by each player, and twenty an hour thereafter. A real chess player can go out only a few plies for a limited number of variations. Chess masters can go out ten or fifteen plies in some situations, which is why they seem to sit there forever between moves. (At one point during one of the games in his match with Fischer, Spassky contemplated his next move for sixty-three minutes! For sixty-three minutes he sat there with his eyes closed. It apparently exhausted him, because after having made his move he then lost the game.) Next our model gives us rules as to the consequences of choosing a given move. In the model the rules are very simple: win, lose, or draw. In a real game they are terribly complicated and involve things like control of squares as well as the loss of pieces. Finally there must be a method of quantifying the "score" for each of the choices—the minimaxing—so that the machine can choose the move that offers the best score. That, in the large, is how to make a chess machine.

Shannon did not actually make one. That appears to have been done first by the British mathematical genius Alan Turing. As in nearly everything that Turing did—he committed suicide in 1954 by eating a poisoned apple—there was something oddly moving about this "machine" which he created in 1950. He acted as his own computer. That is, he developed a chess-playing program following Shannon's outline—actually improving on it—

and then proceeded to try to carry out its instructions himself, making the computation before each move and then moving accordingly. This was sufficiently taxing so that he played only one game—against a relatively weak human player who did not know the program. In the process Turing made a mistake in his minimax computations. At every ply Turing considered all the possible continuations except for the ones he "knew" would lead to nothing interesting, and in one case he was wrong. To keep the number of computations at all feasible, Turing considered only plies that stopped at what he called "dead" positions—positions in which, for example, there is an exchange of pieces and the rest of the material is static. It took Turing many minutes of computation between moves, and it is reported that the "machine" program was rather aimless. It lost, but it was a beginning.

For the next several years the development of chess-playing programs followed closely the development of the computing machines themselves. Here we must make a distinction. Following the brilliant postwar work of John von Neumann, the logical architecture of so-called all-purpose serial processing computers was established. A serial machine processes one command after another. Deep Blue, in contrast, does parallel processing in which various computational procedures take place simultaneously. The von Neumann machines are "universal" in the sense that they follow programs in the same way whether the programs are designed to play chess or do word processing or solve equations. Personal computers, for example, are universal—at least if you restrict yourself to an operating system such as DOS or Windows. You don't need to buy a new machine to run a chess program. You just take it home from the store and load it onto your laptop. But the really good chess machines like Deep Blue are "dedicated." They are designed specifically to play chess, and if they do anything else at all they probably do it badly.

In the 1950s there were very few computers that were fast enough to do anything interesting when it came to playing chess. There was the MANIAC at Los Alamos and later the JOHNNIAC—

named after von Neumann—at the Rand Corporation, both of which spent most of their time doing computations relevant to nuclear weapons design. The fastest machine seems to have been the IBM 704, which could carry out 42,000 operations a second—an absurdly small number by today's standards. Each of these machines had a chess program. The first one was designed for the MANIAC, which could carry out 11,000 operations a second. This was too slow for playing a real chess game, so the board was reduced to 6 x 6 squares as opposed to 8 x 8. In this configuration the MANIAC was able to examine all continuations to a depth of two plies, though it took some twelve minutes a move to do so. The first real 8 x 8 chess-playing program was designed in 1957 by a group at IBM led by Alex Bernstein (no relation), who was himself a strong chess player. In designing this program, criteria were introduced to limit the number of moves considered at each ply. As every chess player knows, in any given position there are always many things one could do according to the rules but which are manifestly absurd. There is no point in working out their consequences since a glance at the board shows that they will lead nowhere. Of course, built into this "glance" are all sorts of rules that a chess player could probably articulate if pressed. In a machine program these rules must be articulated and built into the program. In Bernstein's program, at any given position only seven plausible moves were considered—seven selected by the built-in criteria—and these were examined out to two plies. This meant that about 2,500 continuations were actually considered out of a possible 800,000 legal ones. The program took twelve minutes per move. Bernstein, who was a very strong player, had no trouble beating it.

The first chess machine that I played was dedicated, all right. It was called the Fidelity Chess Challenger. It indicated what it wanted to do by flashing lights in the various squares on the board. It had, as I recall, four possible settings: beginner, intermediate, advanced, and something called postal chess. In the latter mode it took forever to make a move—hours—and I never had the patience to play a complete postal game. In any of the settings, if you played white you selected the opening. Then the machine would respond by choosing, presumably randomly, from its menu of defenses appropriate to your opening move. In the course of experimenting with it I came across an absolutely remarkable "bug." I set it to "advanced," and I opened by moving my king's pawn two squares—a standard opening. It followed by doing the same. I then moved my king's knight out to where I could threaten its pawn. It did the same. I took its pawn and it took mine. I then moved my queen in front of my king so that it threatened the knight. It then committed suicide. It moved the knight away, leaving itself open to a catastrophe. This in itself did not surprise me much. The machine wasn't very good, and I could readily beat it in all of its settings. I repeated the sequence several times, but it never learned from experience. I then had the inspired idea of trying the same sequence in its other two settings—"beginner" and "intermediate." In "beginner" it once again committed suicide—no surprise—but in "intermediate" it didn't! It made an appropriate defensive move. This was a surprise. Finally I came up with an explanation: in "beginner" it was too stupid to know that it was in mortal danger, while in "advanced" it was too neurotic to care.

Some years later I had a chance to discuss this with Marvin Minsky, and old friend and one of the pioneers in research into "artificial intelligence." He told me that in 1965 the artificial intelligence laboratory at MIT, of which he was one of the founders, got a Digital Equipment Company PDP-6 computer—a fast computer for its day—to use in its work. One of the "hackers"—a term that Minsky seems to have invented—named Richard Greenblatt programmed it to play chess. The program was pretty good, and the machine actually won tournaments in the novice class. Like Bernstein's program, this one was designed to look at the seven best moves in depth. This at least is what Greenblatt thought it was doing. But Greenblatt introduced a hack, in this case a parasite program that could follow and report on what the main program was doing. The parasite program discovered that because of some programming error, the chess program was analyzing only the six best moves along with the worst possible move. This did not show up for more than a year. The explanation was that after analyzing all seven moves, the machine never chose the worst one. As Minsky put it, ". . . Here you had a sort of demon inside the machine—a sort of self destructive impulse of the worst kind. But it was always censored before it reached the machine's consciousness." He also told me of an even more remarkable bug in a checkers-playing program that had been designed by Arthur Samuel of IBM. Somewhere in writing the program Samuel had gotten a sign wrong. Hence his program was designed to give away checkers as fast as possible. Nonetheless it played excellent checkers! The reason, Samuel decided, was that designing a program to give away checkers rapidly—forcing opponents to take them—involves the same kind of strategy as winning one's opponent's checkers. Whether there is any larger moral in this is unclear.

The next chess machine I played was an entirely different kettle of fish. At the time—1983—it was the world-champion machine, the Deep Blue of its day. It was called BELLE and had been designed and assembled at the AT&T Bell Laboratories. Before I describe the machine and its creators, let me explain just how good it was. The International Chess Federation developed a numerical rating system for chess players. To get such a rating, one had to compete in tournaments. On this scale Bobby Fischer rated 2780. Grand masters had a rating that was at least 2450. The best commercially available chess program at the time rated about 1700. The average player in the United States was rated at about 1400. BELLE was rated at 2200, which meant that it could beat ninety-nine out of a hundred players in the world. In the fall of 1983 the International Chess Federation gave it an official master rating. Clearly Deep Blue must have a grand master rating. Not having played in chess tournaments, I have no exact idea of what my rating is, but BELLE was certainly the strongest player I have ever played.

The reason I got to play it was that at the time I was gathering material for articles and ultimately a book—*Three Degrees Above Zero*—about the Bell Labs. I had read about BELLE and therefore knew that I had at least an indirect connection with one of its creators, Joseph Condon, a physicist at the Laboratories. When I was a graduate student, as a "punishment" for flunking an oral examination I was required to take a course in experimental modern physics. I was absolutely hopeless in the laboratory, but, as luck would have it, I drew as my partner Paul Condon, Joe's older brother. They are both sons of the late E. U. Condon, a very distinguished American theoretical physicist. Paul was wonderful in the laboratory and allowed me to piggyback on his excellent lab reports so that I passed the course—barely. When I contacted Joe at the Labs I told him about this, and he in turn invited me to play BELLE. He also explained something of the nature and origins of the machine.

In 1973 he and his colleague Ken Thompson began writing chess programs. Thompson, who also witnessed my encounter with BELLE, is something of a legend in the computer world. (He was, by the way, one of the referees for Kasparov's match with Deep Blue.) He and another Bell Labs engineer, D. M. Ritchie, designed what is known as the UNIX operating system. If you use the Internet, or almost any other computer network, the chances are that you are speaking UNIX without knowing it. The reason that you don't know it is that you are probably using a "UNIX front end," a program that speaks to you in English or icons and then translates into UNIX. This is all to the good since UNIX commands are notoriously nonintuitive. Condon and Thompson's first chess program ran on a standard all-purpose machine and could evaluate only two hundred positions a second. To improve matters they decided to construct a dedicated chess machine.

Their first one used only twenty-five computer chips, and the results were not much better than the original program. But by now they had become hooked on the problem, which is somewhat strange since neither of them was especially interested in chess. Their next version had 325 chips and could evaluate about 5,000 positions a second. BELLE had some 1,700 chips and could evaluate 160,000 positions a second, substantially fewer than Deep Blue but still pretty impressive. The number of plies that it searched could be adjusted. In fact Condon and Thompson did an interesting experiment by having the machine play itself with, say, white searching four plies and black searching three. They concluded that an increase in ply of one improved the machine's rating by about 250 points—but then it took a fivefold increase in computing time. I was curious as to how the machine would do if it was confronted with some of the great classical positions you find in the chess literature. An example that occurred to me took place in a game that Bobby Fischer had played at age thirteen against the American grand master Donald Byrne. On the seventeenth move—very early—Fischer sacrificed his queen, something that caught Byrne and everyone else watching the game completely by surprise. Fourteen moves later Byrne found himself inexorably checkmated, something that Bobby had seen from the beginning. This persuaded most observers that Fischer was the genuine article. I asked Thompson if they had ever set BELLE up in Fischer's position to see what would happen. He said they had. In nineteen seconds the machine sacrificed its queen and went on to win the game.

To understand in what sense BELLE—and, indeed, Deep Blue—are "dedicated," this is the sort of thing that BELLE did when it evaluated a king position. The evolution of the king was tracked by a series of so-called accumulators so that at any given point in the game the machine knew where the king was. Four special registers denoted the presence of "friendly" pawns near the king. Pawns were assigned a numerical value on a scale in which the queen was given nine hundred and a rook five hundred. But a pawn could get a special "bonus" if it was in the center of the board or had been advanced to where it might eventually become a queen. Knights and bishops that refused to move from their original squares were "penalized"—some amount was subtracted from their original value. If the king was behind its pawns, this structure was also evaluated.

A curious feature of machine chess play is that machines—at least of the conventional type—seem to have difficulty with the "end game" in which there are only kings and pawns and possibly bishops or knights. This is odd because that is one aspect of play in which numerical evaluation plays a large role. If you have ever watched high-level chess players in action in an end-game situation, you can see them count how many moves it takes to, say, queen a pawn as opposed to some defensive activity on the part of their opponent. It really comes down to numbers. But with the machine, perhaps it is the fact that there is so much open territory for it to scan that makes difficulties for it. In any event, BELLE had special hardware that enabled it to cope with end games, though this was not one of its strong suits. In addition, something was added that Condon referred to as the "contempt factor." This was an estimate of an opponent's ability. If Condon and Thompson did not think highly of a prospective opponent, they would raise the contempt factor so that BELLE would not play for a draw even if the conditions warranted. This ultimately led to BELLE's downfall in a match that took place with a CRAY super computer. They underestimated the CRAY, and BELLE refused to play for a draw and lost.

The machine itself, which could fit into a 48 x 48 x 71 centimeter box and weighed about 130 pounds—as Thompson once wrote, "It is portable, but one has to be dedicated to take it anywhere"—sat in a corner of Condon's crowded laboratory. My appointment to play it was for late in the afternoon—Thompson's "morning" since he worked at night and wanted to watch the encounter. We decided that we would play at tournament speed but that we would not use clocks. I felt I would be under enough pressure without the added burden of remembering to start and stop my clock. I chose to play black, which surprised Thompson since, as he said, choosing black drops you about twenty-seven points in rating. But in my limited study of chess books, the one sequence of moves that I had absorbed with any depth was the so-called French defense, a response to a king's pawn opening. I felt that if I played by the book I would at least get some moves in before being totally crushed. BELLE chose to open with its king pawn so that I was able to use my beloved French defense. Thompson told me that this had been a semi-random decision by BELLE, which had been programmed with a number of book openings from which it could choose.

As the game worked, I would make a move which Thompson would then type into the machine. After what appeared to be a millisecond, the machine made its next move, which Thompson would place on the board. Like the Kasparov–Deep Blue match, we used a standard chess board and standard pieces in actually playing the game. (There were no flags.) What I did not know until later, fortunately, was that a parasite program was running simultaneously with BELLE's move generators that offered a running evaluation of my play—the sort of thing you see in chess magazines, with exclamations and question marks placed after moves, except that this was going on in real time. Condon and Thompson never let on, until later, what BELLE thought of me. Actually it gave me fairly good marks—at least in the beginning. After several moves I began to try to put pressure on BELLE's king side. On its twentieth move it castled directly into the pressure. This caused Thompson, who to that point had said almost nothing, to remark, "That's strange." This kind of unnerved me. Had BELLE made a mistake? Was I about to become one of the few people actually to beat it? On the twenty-second move we exchanged queens, and then the roof fell in. First I lost a pawn and then a rook, and on the twenty-seventh move I gave up. There was no comment from BELLE.

In 1950 Alan Turing published an article in the British journal *Mind* entitled "Computing Machinery and Intelligence." It remains one of the most influential articles ever written about computers. The first sentence reads, "I propose to consider the question, 'Can machines think?' " Turing realized that his question could rapidly lead one into a hopeless swamp of semantics. What is "thinking"? What is a "machine"? and the like. So, in the first place, he limited his class of machines to digital computers, which might or might not be electronic. These computers are examples of what Turing called "discrete state machines." The "state" of such a machine might be specified by noting, for example, which switches were on or off, or which spaces in its memory were empty or full. When subject to some input—an electric pulse, for example—the machine makes a jump to another well-defined state. This, in gross terms, is what an electronic computer does. Turing argued that these machines are "universal" in the sense that a digital computer can mimic the activity of any discrete state machine. Hence Turing rephrased his question to ask if there was any imaginable digital computer that "thinks."

To avoid the issue of what "thought" is, Turing invented a game. He called it the "imitation game," though now it is commonly called the "Turing game." Imagine that you are in a room that is divided in two. You are in one half, and you are separated from the other half by a curtain that conceals what is behind it. Nonetheless messages can be exchanged across the curtain. You can carry on a dialogue with whatever is on the other side. Turing suggested how such a dialogue might go:

Q: Please write me a sonnet on the subject of the Forth Bridge.

A: Count me out on this one. I never could write poetry.

Q: Add 34,957 to 70,764.

A: (Pause about thirty seconds and then give as an answer) 105,621.

Which, incidentally, is wrong! But the idea is clear. You keep asking these questions, or whatever, as long as you like. Meanwhile, on the other side of the curtain, they are being answered alternatively by a person—who by definition "thinks"—and by a digital computer. If at the end of the day you cannot tell whether it is the person or the machine that answered any given question then, Turing argued, the machine "thinks."

Of course there is no computer—Kubrick's HAL excepted—that can successfully play the Turing game. But in the narrow domain of chess we may have reached that point. In an interview Kasparov once remarked, "I believe signs of intelligence can be found in the net result, not in the way the result is achieved. If I look at a position and tell you, 'I will go this way,' and the machine looks at the position and comes to the same conclusion, and in many tests like this we reach the same conclusion, I don't care how the machine gets there. It feels like thinking." This was said before his encounter with Deep Blue. What are we to make of this? One could take the position that just as we do not expect a human runner to compete with a Formula One racing car, we should not expect humans to compete with chess machines which can be improved without any seeming limit. Whether Kasparov comes back to beat Deep Blue hardly matters. Just as this version of Deep Blue was qualitatively better than the one that Kasparov beat in 1996, the next one will be still better. We will soon have to acknowledge that a machine will be stronger than any chess master who ever lived or who can ever live. I don't know how that makes you feel, but it makes me feel a little seasick. Chess at its best is, or was, akin to art. The magister ludi was for those of us who love the game a figure of awe. I will never forget the sight of Samuel Reshevsky—a great chess master of a few years ago—taking on a group of some forty of us, which included Marcel Duchamp, in an exhibition of simultaneous chess. There Reshevsky was in person, short and bald, a living human being. Now we have an inanimate contraption that can fit into a large box that will soon be able to make us all look absurd.

Tom Stoppard's Quantum

In the fall of 1994 I received a phone call from the editor of the *New Theater Review*. She informed me that Lincoln Center was bringing in a play by Tom Stoppard called *Hapgood*. The play, she said, was somehow related to the quantum theory, a subject not likely to be fully understood by all of the patrons of Lincoln Center. Could I, she asked, read the play, which had first been performed in England in 1988 and had since been published, and perhaps write a few clarifying words for the *Review*? Although she could not have known this, I had actually written a fair amount about the theater for the *New Yorker*. I had done this under the protective anonymity of "Talk of the Town," so no one knew that it was me. This apart, I had already read in the newspapers that Stoppard had written a play somehow inspired by the quantum theory. I was, and am, a Stoppardian of the deepest dye. I think I have read or seen all of his plays. If any other playwright had said that he or she had taken the quantum theory for their muse, I would have run for the hills. But this was Stoppard!

This was a man who had been inspired to write *Travesties*—a marvelous play—by having read about reciprocal lawsuits involving James Joyce and a minor British consular official named Henry Carr in Zurich in 1918, dealing with financial differences over the production by the English Players of Oscar Wilde's play *The Importance of Being Earnest.* Among other things, Carr wanted Joyce to give him 150 Swiss francs for having bought a new suit which he claimed was a "costume." Anyone who could use this as the basis of a play was, I felt, capable of anything. I agreed to take on the job.

The first thing I did was to buy a copy of *Hapgood.* Upon opening it, two things struck me. First, the original British cast, which included such people as Felicity Kendal, Nigel Hawthorne, and Roger Rees, was remarkable—a real A-level production. Second, in this edition Stoppard had preceded the actual play with an introductory quotation by Richard Feynman about the quantum theory. This looked promising, except that most of the quotation had been elided out—replaced by three dots. Indeed, what had been elided was that part of the quotation that actually dealt with the theory. (I will come back to Feynman's quotation later. It is even more peculiar than I first realized.) Upon reading the play—which is in fact a spy story—I discovered that the physics is represented by a character named Kerner, a physicist, who is a double or triple agent. He has invented an anti-ballistic missile system that uses anti-matter. Lots of luck! He is given to delivering mini-lectures on physics. These are what caught my eye. They are not quite right. Indeed, in places they are quite wrong, or at least quite misleading. I had a brief thought that maybe Stoppard really did understand the physics but had invented the brilliant conceit of having a physicist character who didn't quite. I decided this was too Stoppardian, even for Stoppard.

After reading *Hapgood* I concluded that there were a couple of useful things I could try to do in a brief article. I think I had about a thousand words. I could try to explain that Stoppard's representation of the quantum theory should not be taken too literally, and then I could ask the question, does it matter? After all, Stoppard was writing a *play*, not a textbook on the quantum theory. It was only after my article appeared that I discovered that Stoppard had discussed this very point in an article he called "Matter of Metaphor" which preceded mine in the *New Theater Review*. He explained that he is "not even a closet scientist," having gotten his notion of the quantum theory from a "dozen" popular books. Since he does not say which ones they were, I cannot comment on his sources. But he insisted that he was using science as "metaphor." "*Hapgood*," he wrote, "is not 'about' physics, it's about dualities. No—let the playwright correct the critic in me—*Hapgood* is not about dualities, of course, it's about a woman called Hapgood and what happened to her between Wednesday morning and Saturday afternoon in 1989 [this date is peculiar since the play was first performed in 1988, but there it is] just before the Berlin wall was breached." Having said this, in the next paragraph Stoppard seems to take it back. He goes on, "As a matter of fact, the story has much more to do with espionage than physics, but I won't deflect any compliments that might be going for a play with a reasonably plausible physicist on board, because the springs of the play are indeed science; it was only in looking around for a real world that might express the metaphor, that I hit upon the le Carré world of agents and double agents. The physics came first, the woman called Hapgood came second."

When I read this, my thought was that Stoppard had made a preemptive strike on me—except that, as far as I was concerned, Kerner, his physicist, was not "reasonably plausible," unless you think a reasonably plausible physicist talking to lay people, which is what Kerner does in the play, would say things about physics that would be superficially understandable but not quite right. Indeed, this was one of the things I had wanted to discuss in my article, namely, if you use science as metaphor, does it matter if you don't get it quite right? Does it matter if there is no "real world"—a world of espionage agents, for example—that correctly expresses what quantum theory is actually about?

Here is something of what I had in mind. *Travesties* is "about" Joyce. Well, it isn't really about Joyce. It is about Zurich. But it isn't exactly about Zurich. It is about people like Joyce and Lenin who found themselves in Zurich at the same time—chance. Joyce is, if you like, a metaphor, and so is Lenin. Nonetheless there are limits. If Stoppard had had his metaphorical Joyce as the author of *War and Peace* or his metaphorical Lenin as the author of *The Leviathan,* the play would have been destroyed. One would be constantly saying, "But Joyce didn't write *War and Peace.*" Since the whole audience presumably knows this, the metaphor would not work. But in *Hapgood* we are in a different situation. I and perhaps six other physicists who saw the play know that what ultimately troubled Einstein about the quantum theory was not God throwing dice (I will come back to this later) but something much deeper and more complex which still bothers a great many people, even today. Thus when Stoppard's Kerner says that what concerned Einstein was that quantum mechanics "finally made everything random . . . a God who threw dice," I react to this in much the same way I would have done if Stoppard's metaphorical Joyce had taken credit for *War and Peace.* The play stops working for me. In the article I wrote I treaded on all of this very lightly. I said that I was pleased that Stoppard had taken quantum mechanics as his muse but noted that his use of it reminded me of a friend who had taken up drawing in middle age. He drew a bird and then remarked, "This looks like a bird, but no bird looks like this." In the article I felt that I had not really explained adequately what my concerns were. There wasn't space, and it was the wrong forum. So I am returning to it here.

I would like to begin by discussing Einstein's attitudes—note the plural—toward the quantum theory. If anyone is a metaphor for the theory, it is Einstein. I see in Einstein at least six different attitudes. This is perhaps not too surprising since from 1905—

the date of Einstein's first paper on the quantum—until his death a half-century later in April 1955, Einstein thought incessantly about the quantum theory. He never made peace with it, but not for the reasons that Stoppard's Kerner believes. Let me try to trace the evolution of Einstein's ideas—one could easily write an entire book about them.

At the turn of the century Max Planck, Einstein's somewhat older countryman, posited a mathematical formula that described how the wavelengths of radiation in a heated cavity are distributed in intensity—at what wavelength, for example, the radiation is most intense. It turns out that this depends only on the temperature of the walls of the cavity and not on what material the cavity is made of. Planck came upon his formula, after several unsuccessful tries, by a mixture of inspired guesswork and extrapolation from experiment. Naturally he wanted to derive his formula from basic physics, that is, the classical physics of the nineteenth century. Here he got stuck. He found that the only way he could derive what we now call the "Planck distribution" was to assume that the walls of the cavity could emit and absorb radiation only in discrete packets of energy—"quanta," he called them. My first great teacher in physics, the Austrian-born physicist-philosopher Philipp Frank, explained this to us in terms of beer. Radiant energy from the wall could only be "bought" and "sold" in "pints" and "quarts." Nothing in classical physics provided for a limitation like this, so Planck spent the next decade trying, unsuccessfully, to find an alternate derivation.

In 1905 Einstein adopted a completely different attitude—the first of his attitudes. He decided that deriving Planck's formula from classical physics was impossible. So he chose simply to accept the formula as true and see what it implied. In explaining Einstein's discovery I will use anachronistic language because that is the language we use now and that is the language that Stoppard's Kerner uses when he tries to explain this to a British agent named Blair. The language I will use involves the terms "particle" and "wave." In his 1905 paper Einstein does not refer to light as consisting of particles. He says rather that the energy in a beam of light is made up of a "finite number of energy quanta, localized at points of space, which move without subdividing and which are absorbed and emitted only in units." Only some twenty years later, when it became well established that the quanta carried momenta as well as energy, did they come to be referred to as "particles." In any event, when Einstein studied Planck's formula he realized that the part of it that referred to the long wavelengths—which had been derived from classical physics—represented the classical picture of light propagating as waves. But he discovered that the short wavelengths did not. These he found behaved like discrete packets of energy. To return to Professor Frank's image, not only was the "beer" bought and sold in pints and quarts, but in the barrel it was *always* arranged in pints and quarts. Einstein did not try to "explain" this—explain in terms of what? He concentrated on seeing if it had any predictive consequences. He found that it did. It gave a good account of what had been observed about the liberation of electrons from the surface of a metal when light is shone upon it—the so-called "photoelectric effect" which is now commonly used to open doors. It was for this work, and not relativity, that Einstein was awarded the 1921 Nobel Prize in Physics.

This is what Kerner attempts to explain to Blair. His explanation is not bad except that, as many people do when they try to deal with this, at the end he falls into a linguistic trap. He says, "There is no explanation in classical physics. Somehow light is particle and wave . . ."—Stoppard's duality. But by using this language he has erected a "paradox." We know what a "particle" is—something like a billiard ball—and we know what a wave is—go to the beach. How then can something be "particle and wave"? This is, of course, the wrong way round. Light is light is light. It is *neither* a particle *nor* a wave. Its behavior is only "paradoxical" when we try to describe it in the limited language of particles and waves. When these features about nature began to emerge, someone made a waggish suggestion that these entities be called "wavicles," a name that fortunately did not stick, but at base the idea was a good one—a new name. In physics we have enough trouble without enmeshing ourselves in anachronistic linguistic fabulations. To say that the nature of light is "paradoxical" is something like saying that the nature of fish is paradoxical because they swim under water. We do much better when we actually study how they swim.

Hence Einstein's first attitude toward the quantum was to accept it as a fact and to examine how it "swam." But this is not a theory. If we believe—as Einstein did profoundly—that the inner workings of nature are accessible to the human intellect, then there had to be a theory, a quantum theory. While Einstein thought constantly about such a theory, the next real breakthrough came in 1912, at the hands of the Danish physicist Niels Bohr. This had to do with the structure of the atom. In 1910 Bohr's mentor (he had studied with him in Manchester), the New Zealand–born experimental physicist Ernest Rutherford, had discovered the atomic nucleus. This is the massive, tiny, positively charged core of the atom. Around it circulate the much less massive negatively charged electrons. The simplest atom of them all, hydrogen, has a positively charged nucleus which we call the proton and a negatively charged electron that circulates about it. But two things—they are related—are wrong with this picture. When it accelerates the electron radiates and hence loses energy. With this loss of energy the electronic orbits should collapse and the electrons fall into the nucleus. Matter would be unstable. Moreover, there would be no special pattern to this radiation. It would consist of a jumble of wavelengths. But in fact the spectrum of light that emanates from an atom like hydrogen consists of lines whose wavelengths are related to one another by elegant and rather simple mathematical expressions. How does this happen? Bohr solved both problems at once by assuming that the electronic orbits were themselves "quantized." You couldn't have any old orbit but just Bohr's special orbits. When an electron made a transition from a higher to a lower orbit—a quantum leap—radiation was emitted whose energy corresponded to the energy difference the electrons had in each of the two orbits. The orbit of least energy, the so-called "ground state," was absolutely stable, so matter did not collapse. Using this picture, Bohr was able to derive the mathematical formula for the spectrum of hydrogen.

Einstein's attitude toward Bohr's discovery was quite simple. He thought it was marvelous. He developed an admiration and respect for Bohr that lasted a lifetime, even though they came to disagree totally on the quantum theory. For the next decade most theoretical physicists devoted themselves to adumbrating Bohr's discovery. The results became known later as the "old" quantum theory, to be distinguished from the revolution that began in 1925, which produced the "new" quantum theory or, simply, *the* quantum theory. Einstein's participation in the "old" quantum theory was limited, though in 1917 he published a paper on the nature of radiation which decades later became the principle of the laser. The focus of Einstein's work was rather on his newly created theory of gravitation. I doubt that any of the physicists who worked on the "old" quantum theory thought it was the final answer, or even close. There were too many loose ends. The theory, which was an uneasy mixture of classical physics (orbits) and quantum physics (quantized orbits), could not account for why one line in an atomic spectrum was more intense than another. This was related to the question of what determined how often the electrons, once elevated into the higher Bohr orbits, jumped back down into the ground state. In fact, what were these quantum leaps? Did *they* have orbits? Even if one was willing to put these fundamental questions aside and simply work with Bohr's postulates, one could not account for the spectra of atoms with several electrons. There had to be more to it.

The first clue came from a totally unexpected source. In 1923 Prince Louis de Broglie, who belonged to an ancient French family, presented his Ph.D. thesis to his adviser, Paul Langevin. (Langevin, incidentally, a decade earlier had had a scandalous love affair with Marie Curie and had survived a duel.) Langevin was not quite sure what to make of de Broglie's proposed thesis, so he sent it to Einstein to evaluate. De Broglie's idea was that electrons, which had been considered as "particles," could also have a wavelike nature that could manifest itself under appropriate circumstances. In other words, like light, the electron was also a "wavicle." His argument for this was based largely on symmetry, though he realized that this picture gave a neat accounting for Bohr's quantized orbits. The allowed orbits, he argued, were just those in which a whole number of electron wavelengths fitted together around the orbit with no overlap. The wavelengths of these electrons, in, say, atoms, were predicted to be something like a thousand times smaller than those of visible light, which explained why ordinary experiments on electrons had not heretofore revealed these wave characteristics.

Einstein was very interested in de Broglie's idea. He may have had similar ideas. He liked the fact that de Broglie's conjecture could be tested. He wrote to Hendrik Lorentz (the Dutch physicist whom Einstein used to say was, of all the physicists he had known, the one he admired most): "The younger brother of . . . de Broglie [Louis de Broglie had an older brother, Maurice, who had become a distinguished x-ray spectroscopist, work that he carried out in a laboratory in the family mansion in Paris] has undertaken a very interesting attempt to interpret the Bohr-Sommerfeld quantum rules. [This refers to the matter of fitting a whole number of wavelengths around a Bohr orbit. Arnold Sommerfeld was a distinguished German theoretical physicist who had done as much as anyone to develop the "old" quantum theory] . . . I believe it is a first feeble ray of light on this the worst of our physics enigmas. I, too, have found something which speaks for his construction." In 1927 C. Davisson and L. Germer in the United States and, independently, G. P. Thomson in Britain carried out experiments that directly confirmed the wave nature of electrons. The idea was to use a so-called diffraction grating—basically a series of regularly shaped openings in matter—and allow a beam of electrons to pass through it. Diffraction gratings were a familiar way, from the beginning of the nineteenth century, to demonstrate the wave nature of light. If light was a wave, the light waves from the different closely spaced openings would "interfere" with one another. When two light waves interfere, the resulting wave can be reinforced at the places where the waves interfere "constructively"—large amplitudes of one wave add to the large amplitudes of the other—or the two waves can negate each other at those places where the amplitudes are out of phase—"destructive interference." This produces a series of dark and light patterns on whatever light-collector is placed behind the diffraction grating. By measuring these patterns, the wavelength of the light that produced them can be determined. This is exactly what was done for the electrons. But since the wavelengths were so much smaller, the "diffraction gratings" themselves had to be on the atomic scale. Davisson and Germer used atomic crystals. The results were a stunning confirmation of de Broglie's conjecture.

Even before the experiments, the question arose as to what these waves were. Were the waves the electron? Or was the electron something to which the waves were attached and somehow guided the electron? This seemed to be Einstein's original idea. He began referring to the waves as "Führungsfelder"—"guiding fields." Whatever they were, de Broglie had not supplied any kind of equation that governed their propagation. But this was done in a series of magnificent papers, beginning in 1926, by the Austrian-born theoretical physicist and general polymath Erwin Schrödinger. When Einstein saw the first one he wrote to Schrödinger, "The idea of your article shows real genius." Schrödinger's equation enabled one to calculate how these "guiding fields" developed in time. At first it looked to Einstein as if a visualizable, deterministic theory was at hand. There was at least one critical voice. That came from the then young German physicist Werner Heisenberg. He did not precisely like this seemingly classical visualizability. He invented what appeared to be a different version of the new atomic mechanics in which the waves did not play a role. One could use his so-called "matrix mechanics" or, alternatively, the Schrödinger equation to derive such things as the energy levels of the hydrogen atom. In fact the British theoretical physicist P. A. M. Dirac showed that the two theories were really one and the same—simply different representations of an underlying theory which came to be called quantum mechanics. It was a matter of convenience which representation one used to treat a given problem.

To understand the next step, the Schrödinger wave picture is the most convenient. One of the most significant facts about the electron is that when you observe it, it appears to be localized, the way one would imagine a particle should be. Either it is in an atom or it is at a precise location in a detector. Waves, on the other hand, might appear to be objects that, loosely speaking, are spread out over space. If one associates a wave with an electron, how then do you make it localizable? This is actually not so difficult. You add many waves together to form a "packet," and you choose them so that their collective amplitude is large only in a small region of space. You take advantage of the fact that waves can add up, or subtract, depending on how they overlap. If you try to do this you will learn that the smaller the dimension of the packet, the more waves you have to add together. But these waves carry energy and hence momentum. So another way to state this observation about the localization of wave packets is that the more they are compacted in space, the wider the variety of momenta that will play a role in the waves making up the packet. A reader who has some familiarity with the Heisenberg uncertainty principles may have just heard a bell go off. We see in the wave packet just the sort of antimony that Heisenberg soon realized was a deep general feature of quantum mechanics. The more we try to make the electron into a localized wave packet in space, the more dispersion there is in the momenta that make up the packet. The more precise the localization in space, the less precise the momentum.

But this dispersion of the momenta leads to something else, something that led to the downfall of the last remnants of classical physics in the quantum theory. The Schrödinger wave equation tells us how these wave packets propagate in time. In this sense quantum mechanics is fully deterministic. If you specify the wave function at some point in space and time, and describe the forces, the Schrödinger equation will tell you what the wave packet looks like in the future at any point in space, at any time you specify. But in the case of the highly localized packet, it may tell you something you don't want to hear. Suppose at some moment in time you create a wave packet that is localized, say, to the size of an atom. Then you ask what happens. What happens is that the wave packet begins to disperse in space. One way you might say this is that the different momenta it contains begin to pull it apart. In a few days you will begin to find traces of it at the edge of the solar system! How can this possibly be consistent with localizability? Enter Max Born.

Max Born, who died in 1970, was a few years younger than Einstein. The two of them—really three if you count Born's wife Hedwig—were very close friends, especially in the early years and especially until Born's disagreements with Einstein about the quantum theory drove Born into a state of near despair. We know a lot about this because, the year after Born died, the letters exchanged among the three of them over forty years were published. One of them, which I will present shortly, describes Einstein's attitude toward the work of Born and his younger colleagues such as Heisenberg as it was developing. It was Born who first understood what the Schrödinger wave packet meant. The four-page paper that Born published in June 1926 is certainly one of the most important in the physics of the twentieth century. Born suggested that the Schrödinger-de Broglie waves were waves of *probability.* What does such a bizarre statement mean? What Born said was that the amplitude of the wave packet—or "wave function," as it is usually called—could be used to compute the probability of, say, the electron's position in space. Where the amplitude was small, it was unlikely that you would find the electron. Where it was large, such as at the positions of the Bohr orbits, you would very likely find the electron if you looked for it in the atom. The spreading out in space of the wave packet meant that in the course of time you had a very tiny chance of finding the electron very far from where it started. But when you found it there it would be localized *there*. There is a there there, and the there is the electron. This is the interpretation of the wave function we have maintained ever since.

What did Einstein think of this? It is important right off to understand that Einstein did not object to the use of probabilities in physics. He was one of the creators of what is known as statistical mechanics. Its very name suggests statistics and probability. It is applied to ensembles consisting of an unthinkably large number of elements—particles if you like. There is no hope in any practical sense of following the destiny of these particles individually. There are too many of them, and their behavior on an individual basis is too erratic to be interesting. Statistical mechanics describes the average behavior of the ensemble. It will tell you, for example, the average speed of a molecule of a gas with some given measured temperature. It will also tell you how big a deviation, on average, you are likely to find *from* the average. Just as there is some probability—infinitesimal—that an electron with an initially tightly confined wave packet will be found later in outer space, so there is some infinitesimal probability that one of the gas molecules will be found to be moving at speeds close to that of light, even if the average molecular speed is, say, that of an airplane. This is not what bothered Einstein. What bothered him was that people like Born, Heisenberg, Bohr, and the rest thought that, unlike the classical situation where there was an underlying deeper theory such as Newtonian mechanics, to which the statistical theory was a useful approximation, nothing underlay the quantum theory. What you saw was what you got.

On December 4, 1926, Einstein wrote a letter to Born, echoes of which have been with us ever since. One of them found its way into Stoppard's play. Here is what the relevant part of the letter said: "Quantum mechanics is certainly imposing. But an inner voice tells me that it is not yet the real thing. The theory says a lot, but does not really bring us any closer to the secret of the 'old one' [Einstein's affectionate way of referring to God]. I, at any rate, am convinced that *He* is not playing at dice. . . ."

I will come later to what Stoppard's Kerner has to say about all of this. His knowledge of Einstein's attitudes seems to have frozen at this point—"Einstein couldn't believe in a God who threw dice," he notes. As I will try to explain in the rest of this essay, there was a lot more to it than that.

For a few years after Einstein wrote this letter, he tried to destroy the new quantum mechanics. He did this by inventing devices—imaginary devices—that he thought contradicted one or more of the various Heisenberg uncertainty principles. I have always thought that in creating these cunning machines he was reaching back to his days in the Swiss National Patent Office in Bern, where he used to make his living by examining inventions like this. As far as I can tell, the last of Einstein's "inventions" was presented to Bohr as a little surprise on the occasion of an international conference held in Brussels in 1930. This invention was an attempt to destroy the Heisenberg uncertainty principle that relates the precision with which an energy can be measured to the time it requires to do the experiment—the more precisely you want to measure the energy, the more time the experiment will require.

To this end, Einstein imagined a box with a clock attached to it. Inside the box there is a radioactive element. The clockwork can open a door in the box at some well-defined time, and out pops some of the decay material. The box sits on a scale, and its weight is measured before and after the emission. This fixes the energy of what has been emitted, and we can apparently make the time interval during which the box door is opened as short as we like, hence seemingly producing a contradiction with Heisenberg. Bohr was momentarily stunned—more than momentarily since he spent a sleepless night over it. By the next morning he realized that Einstein had overlooked an effect from his own theory of gravitation! In a gravitational field a clock runs slower than an identical clock outside it. Hence to know the rate of a clock in a gravitational field that varies from place to place, you must know where you are in it. The clock will have different rates at different places in the field. But Heisenberg's uncertainty principle limits the precision with which the clock's position can be known without imparting some momentum to the scale. This upsets the measurement of the energy. Bohr found that it all worked out just right so that there was no inconsistency. Perhaps Einstein realized that there was something deeply psychological about making a mistake like this. One doesn't know. One does know that he did not try to invent any more examples of this sort. He now adopted what I think was his final attitude toward the theory: it failed to give a "complete" description of reality, or what he perceived to be reality.

To approach this attitude we can once again return to Stoppard-Kerner. What they understand about this, as well as what they apparently do not understand, are both instructive. Let us begin with the quotation from Feynman that Stoppard put as a prelude to the printed edition of his play. In Stoppard's elided version it reads: "We choose to examine a phenomenon which is impossible, *absolutely* impossible, to explain in any classical way, and which has in it the heart of quantum mechanics. In reality it contains the *only* mystery. . . . Any other situation in quantum mechanics, it turns out, can always be explained by saying, 'You remember the case of the experiment with the two holes? It's the same thing.' "

When I first read this quotation as Stoppard had it, I did not pay much attention to it. I had seen quotations like this before and may even have heard Feynman say them. Also, I agreed completely with what he said. But when I decided to have a look at all of this again, I thought it might be interesting to see what the three dots concealed. I was astonished. The part of the quotation that precedes the dots is taken from the opening section of the lectures that Feynman gave to undergraduate physics majors at Cal Tech in the early 1960s. They were later published under the title *Lectures in Physics*. The quote is from Volume III, the lectures on quantum mechanics. I then looked at the rest of the lecture to find the rest of the quote. It isn't there! Rather, it comes fairly near the beginning of the sixth of a series of lectures that Feynman gave at Cornell University in 1964 and which were later published under the title *The Character of Physical Law*. The three dots represent a curious kind of quantum leap from one book to the other. What is elided are the contents of the two lectures which, as I will now try to demonstrate, Kerner does not really comprehend.

He starts fairly well. Somewhere near the beginning of the play he wants to explain the difference, as he understands it, between the nature of light and the nature of, say, a giraffe—his example, which, he says, exemplifies "objective reality." I will give the dialogue that ensues—it is a bit lengthy—and comment as we go along like some sort of Greek chorus. Kerner begins:

KERNER: . . . objective reality is for zoologists. 'Ah yes, definitely a giraffe.' But a double agent is not like a giraffe. [This is a spy story.] A double agent is more like a trick of light. Look. Look at the edge of the shadow. It is straight like the edge of the wall that makes it. Your Isaac Newton saw this and concluded that light was made of little particles. Other people said light is a wave but Isaac Newton said, no, if light was a wave the shadow would bend round the wall like water bends round a stone in the river. [If one wants to be pedantic, this account is wrong on nearly all points. The edge of such a shadow is notoriously fuzzy, which is due precisely to the fact that light bends "round the wall," something that had been observed in the seventeenth century and was then confirmed by Newton's contemporary and nemesis, Robert Hooke. Newton, unlike his followers, was very cautious about saying what light "was." He was much more interested in learning—to use my previous image—how it "swam." But for the purposes of what follows, these points will not be the ones at issue.] Now we will do an experiment together.

BLAIR: *Now,* Joseph?

KERNER: Absolutely. In this experiment you have a machine gun which shoots particles . . . which we call . . . ?

BLAIR: (*tentatively*) Bullets?

KERNER: Bullets. You are shooting at a screen, say like a cinema screen. But between you and the screen I have put a wall of armor plate, so of course none of the bullets get through. Now I open two slits in the armor plates. [Feynman's two holes.] (*He gestures to indicate the two slits a few inches apart.*) Now you shoot many times. Now you stop shooting and examine the screen and naturally it has bullet holes where some of the bullets came through the two slits. Opposite each slit there is a concentration of bullet holes, and maybe just a few holes to left and right from ricochets. This is called particle pattern. If your gun was a torch [flashlight] and light was bullets, as Newton said [sic], this is what you will get. [Actually Newton didn't say anything about this experiment, since it was not conceived and performed—by Thomas Young—until the beginning of the nineteenth century! But again, this is not relevant to the point that Kerner will try to make.] But when you do it [with light] you don't get it. You don't get particle pattern. [Keep in mind that Kerner is talking with a Russian accent, which is reflected in the dialogue.] You get wave pattern. Wave pattern is like stripes—bright dim, bright dim, across the screen. This is because when a wave is pushed through a little gap it spreads out in a semi-circle, so when you have *two* gaps you have two semi-circles spreading out, and on their way to the screen they mix together—so now you understand everything, yes?—no, you understand nothing, OK: where a crest from one lot of waves meets a crest from the other lot of waves you get a specially *big* wave [constructive interference], and where a crest meets a dip, the wave is canceled out [destructive interference]—strong, weak, strong, weak, on the screen it looks like stripes.

BLAIR: Joseph—I want to know if you're ours or theirs, that's all.

KERNER: I'm telling you but you're not listening. So, now light was waves [the "now" here is a little puzzling, but presumably what Kerner is trying to say is that after Thomas Young did his experiments the general consensus was that light was a wave], Isaac Newton [sic] was wrong about bullets. But when light was waves there came a problem with a thing called the photoelectric effect, a real puzzle, which I will describe in detail with historical footnotes—OK, it was just to see your face. But this puzzle was a puzzle because everyone knew that light was waves. [Here Kerner has fallen into the trap we have discussed earlier. The "puzzle" is more semantic than anything else. Light is light is light.] Einstein solved it. Or rather he showed that if light was bullets after all, there was no puzzle. [Light is no more "bullets" than it is waves. In this sense Einstein did not "solve" anything. His work revealed an aspect of light that had theretofore been concealed.] Or rather he showed that if light was bullets after all, there was no puzzle. [What puzzle? The semantic puzzle? The solution to that is to stop using the words "wave" and "bullets," which Kerner is unwilling to do.] But how to explain the stripes on the screen, the wave pattern? Wave pattern happens when light from two slits mixes together, but your particles can't do that—your bullet of light has already hit the screen before the next bullet is even fired. So there is only one solution. [Here Kerner's semantics are about to go off the deep end and take us with them. *His* "particles" can't do that. But who cares what his particles can or cannot do. What we care about is what light actually does!]

BLAIR: What's that?

KERNER: Each bullet goes through both slits.

BLAIR: That's silly. [I wish I had been there to coach Blair. I would have told him to ask Kerner to *prove* it—prove that each "bullet" goes through both slits. But Kerner now goes on in that general direction anyway.]

KERNER: Now we come to the exciting part. We will watch the bullets of light to see which way they go. This is not difficult, the apparatus is simple. So we look carefully and we see the bullets one at a time, and some hit the armor plate and bounce back, and some go through one slit, and some go through the other slit, and, of course, none go through both slits. [In other words, contrary to what Kerner said previously, if we actually set out to prove that a "bullet" goes through both slits, we find that it doesn't. Let us hang on to this. It will become important soon.]

BLAIR: I knew that.

KERNER: You knew that. Now we come to my favorite bit. The wave pattern has disappeared! It has become particle pattern, just like with real machine-gun bullets. [Again, to be pedantic, this is not really correct either. The pattern does not look like "real machine-gun bullets." If we focus on the light that goes through one of the slits, it also exhibits an interference pattern—the phenomenon of "diffraction." This interference pattern is different from the interference pattern that is produced when we don't determine through which slit the light has gone. The striking thing is that these interference patterns are generated in the course of time even if we let the light through one quantum at a time. If we look at this in terms of probabilities and wave functions, we can say that the places where there are many hits on the screen correspond to the places where the Schrödinger probabilities are large. To lapse, just this once, into mathematical language—you will see that it won't hurt—let us call the Schrödinger amplitude produced by the first slit A, and the second B. Then if we observe through which slit the light quantum has passed, the resulting pattern is predicted to be $A^2 + B^2$; and if we don't observe through which slit the light quantum has passed, we must use the expression $(A + B)^2$ to predict the resulting pattern. This is how light behaves, no matter how we may feel about it.]

BLAIR: Why?

KERNER: Because we looked. [I think this brings an irrelevant subjectivity to the situation. It assumes that a certain kind of observation involving a human has been made. But the "we" could be an automated apparatus that functioned while "we" were on vacation. Some of my colleagues unfortunately like to talk about the role of "mind" in all of this. It is my view that in this discussion mind doesn't matter.] So, we do it again, exactly the same except now without looking to see which way the bullets go; and the wave pattern comes back. So we try again while looking, and we get the particle pattern. [So to speak.] Every time we don't look we get wave pattern. Every time we look to see how we get wave pattern, we get particle pattern. The act of observing determines the reality. [The introduction of the term "reality" is the iceberg that is guaranteed to sink any sensible discussion of these matters. Einstein loved this term, and I think his discussions about the quantum theory became less and less sensible. Here we perform two entirely different experiments—observing or not choosing to observe through which slit the light quanta go—and we get two entirely different outcomes. Is that surprising? Couldn't I argue, if I wanted to be perverse, that if I have a bowl of soup and do one experiment to measure its temperature and another to measure its density, the act of observing has determined the "reality" of either the temperature or the density? This is not what caused Einstein's problems. In the case of the soup we would all agree that at all times the soup had both a temperature and a density "reality" which we could measure separately or simultaneously, as we chose. In quantum mechanics we cannot assume that something we have not measured has a "reality." For example, we cannot assume in the case of the two slits that even if we didn't observe it, the photon must have gone through one hole or the other. If we don't observe it, we are not doing a hole-determining experiment, and we have, quantum mechanics insists, no right to speak as if we had done one. *That* is what bothered Einstein. I will come back to this momentarily, but you can see already that it goes much deeper than the fact that the theory uses probabilities.]

BLAIR: How?

KERNER: Nobody knows. Einstein didn't know. I don't know. There is no explanation in classical physics [nor in quantum physics for that matter]. Somehow light is particle and wave. [Light is light is light.] The experimenter makes the choice. You get what you interrogate for. . . .

Kerner uses this as a metaphor for what it means to be a double agent. "You get what you interrogate for." I leave it to the reader to decide if a double agent is more like light or more like soup.

The other excerpt from the play I want to consider occurs somewhat later in the first act. This time it involves Kerner and the play's heroine, Hapgood. As the plot unfolds, it turns out that Hapgood and Kerner have had an affair which produced a son whom she named Joe. At the very end of the play, Joseph and Joe meet for the first time, one gathers. Joseph is about to go off into some kind of exile. But he and Hapgood edge into a metaphorical conversation about physics. I will again quote the relevant parts and again act as a "Greek chorus." In anticipation, let me point out that if the two-hole experiment is done with electrons, or any other so-called elementary particle, the same quantum mechanical results are found. All these objects behave like "wavicles." Now to Kerner.

KERNER: . . . The particle world is the dream world of the intelligence officer. An electron can be here or there at the same moment. [This nonsensical statement persuades me that neither Kerner nor Stoppard actually read what Feynman said behind those three dots. Here is what he says in his quantum mechanics lecture in his book *The Character of Physical Law*. The language is Feynmanian. "The thing" he writes, "(the electron) which is coming (to the detecting screen) in lumps—it has a definite size, and it only comes to one place at a time." Note well—*"One place at a time!!"* In so far as you determine the electron's *place*—its location—it has one and only one. What is spread over space is the probability that the localized electron can be found in some given place. What could be clearer?] You can choose; it can go from here to there without going in between; it can pass through two doors at the same time, or from one door to another by a path which is there for all to see until someone looks, and then the act of looking has made it take a different path. [This is such a gumbo that one despairs trying to sort it out. The crucial point to insist on, and the thing that Einstein really objected to, is that quantum mechanics is consistent only if you do not insist that things that have not been measured have a "reality," a sort of independent existence. This is how the classical world—the world of common sense, the world of soup and billiard balls—works. I have no doubt that if I take a coin and slice it into two layers, and keep the slice with the head and give you the slice with the tail, and you take your half with you to Nepal, when I look at my head you will have a tail and have always had one. But this is not how the quantum world works. If you are concerned that something might pass through "two doors at the same time," you must tell us how you know. You cannot say, "Well, common sense tells me that if the electrons, one after the other, have produced this interference pattern, each electron must have interfered with itself, and therefore it must have come through 'two doors at the same time' because that is what billiard balls would have done." Electrons are electrons are electrons. Their behavior cannot be "explained" in terms of classical billiard-ball physics. We are dealing with something entirely new. Einstein refused to accept that this could be the last word. His intuition, which really was the intuition of a classical physicist, persuaded him that this quantum description was incomplete. There had to be more. But what? He was never able to tell us. Nonetheless the discussions he had with Bohr were, I think, highly valuable. They forced us to understand what we are committing ourselves to when we say that we believe that quantum theory is right.] Its movements cannot be anticipated because it has no reasons. [It is true that the "movements" of single electrons cannot be anticipated, but neither can the movements, in any practical sense, of a single molecule in a gas. In both cases, what can be anticipated is the average behavior. The statement "because it has no reasons" is beyond my understanding.] It defeats surveillance because when you know what it's doing [this is Kerner's somewhat foggy way of saying, when you know its momentum] you can't be certain where it is, and when you know where it is you can't be certain what it's doing: Heisenberg's uncertainty principle; and this is not because you are not looking carefully enough, it is because there is *no such thing* as an electron with a definite position and a definite momentum; you fix one, you lose the other, and it's all done without tricks; it's the real world, it is awake.

This last is a fair rendering of the sense of Heisenberg's uncertainty principle, but by this time the damage has been done. In this respect I turn to my last example, part of a speech by Kerner a bit further on. He is trying to explain to Hapgood Einstein's attitude toward the quantum theory. In this case Kerner is like the severely myopic trying to lead the blind.

KERNER: It upset Einstein very much you know, all that damned quantum jumping, it spoiled his idea of God, which I tell you frankly is the only idea of Einstein I never understood. [You want to bet?] He believed in the same God as Newton, causality, nothing without a reason, but now one thing led to another until causality was dead. [It is tempting to dive into this one as well. The problem is that I can't ask Kerner what he means by "causality." For example, if you ask me what, in general, "causes" neutrons spontaneously to decay into protons and other detritus, I would answer, "The weak force." But if you ask me what caused *that* neutron to decay two minutes ago, I could not tell you since, again, the quantum theory is only concerned with average behavior. If this is what you mean by causality, in that sense "causality" is a casualty. If you are satisfied in predicting average behavior, the quantum theory is as "causal" as Newtonian mechanics.] Quantum mechanics made everything finally random, things can go this way or that way, the mathematics deny certainty, they reveal only probability and chance, and Einstein could not believe in a God who threw dice. He should have come to me, I would have told him, "Listen, Albert, He threw *you*—look around, He never stops. . . ."

I do share one thing with Einstein. I do not believe that the quantum theory is the final answer. That is because I do not think any theory is the final answer. It took 250 years from the time Newton made his mechanical synthesis until Planck invented the quantum. Perhaps it will take another 250 years before the quantum theory finally breaks down—perhaps it won't. I have no idea. I suspect that when that happens it will look so strange that people will be trying futilely—to "explain" it in terms of the quantum theory. Some future Stoppard will look "around for a real world that might express the metaphor," and some future skeptical physicist like myself will write "Toto, I don't think we're in Kansas anymore."

SN-1987A

❦ On the evening of February 23, 1987, Albert Jones of Nelson, New Zealand, as was his custom, went out after dinner to the driveway near his home. In the driveway he had set up a large astronomical telescope of his own construction. Jones was then sixty-seven and had recently retired after working in the repair shop of an automobile company in Nelson. But since 1944 he had spent much of his free time as an amateur astronomer studying "variable stars"—stars whose brightness varies in the course of time. He estimated that he had made about 350,000 variable star observations. He had been called the world's greatest amateur observational astronomer, and that summer he was awarded the Order of the British Empire for his work. But on February 23 he began his star scan by examining a relatively nearby galaxy known as the Large Magellanic Cloud which, at about 9:30 p.m. local time (the time he was looking at it), was quite high in the sky. He noticed nothing unusual.

In view of what was about to happen, it is instructive to

quantify Jones's nonobservation. His home-constructed telescope is what is known as a 12.5-inch "Newtonian reflector." This means that in essence it is a long tube into which light enters, where it is then collected and focused by a mirror—in this case 12.5 inches in diameter. The mirror was one of the few things Jones actually bought that was designed for a telescope. The light is reflected out the side of the telescope by a second mirror and can be examined through an eyepiece. In this case the eyepiece was a World War II bombsight. The telescope was powerful enough to detect stars up to what astronomers call the +13.5 magnitude. Astronomers have traditionally used a magnitude scale that, perversely, is adjusted so that the larger the number, the *weaker* the apparent brightness. Negative numbers are assigned to strong sources. Thus on this logarithmic scale the sun has an apparent magnitude of –26.6, the moon –12.6, and the planet Jupiter, at its brightest, –4.4. Jones could have detected any peculiar star with a magnitude of less than +7.5, and he saw none in the Large Magellanic Cloud that evening.

The next morning, as he often did, Jones set his alarm clock for a predawn hour so that he could continue his observations. By this time the Large Magellanic Cloud was setting, and, as he had already studied it last evening, he passed it by. This, as it turned out, was a pity. If someone had continued to watch for an hour or so after Jones had turned his attention elsewhere, they would have seen the most spectacular astronomical event to occur in our vicinity in the last 383 years. The unprepossessing blue star Sk-69° 202, so designated in a catalogue by Nicholas Sanduleak of Case Western Reserve University, had suddenly become a supernova—SN-1987A—so bright that it would have been visible with the naked eye. While many supernovas are observed each year, this was the first one in nearly four centuries seen to explode in our own or a neighboring galaxy.

The next evening Jones was back at his telescope. Now, to sort out the sequence of events, it is useful to describe things in what astronomers call "Universal Time." (This is also known as

Greenwich mean time.) The advantage of doing this is that the effect of the longitude gets separated out, and one can see at a glance what event is earlier and what is later. In these units Jones's initial nonobservation was recorded at 9:22 a.m. UT (Universal Time) on February 23. It was now 8:52 p.m. UT the following evening. When Jones pointed his telescope at the Large Magellanic Cloud, the supernova, in his words, "popped out" at him. By this time it was +4.4 magnitude, which is to say, it was as bright as a star visible to the naked eye. He at once called Frank Bateson in nearby Touranga.

Bateson had been, for sixty years, the director of the Variable Star section of the Royal Astronomical Society of New Zealand. Jones said to Bateson, "Frank, there is a star in the Large Magellanic Cloud where there was no star before." Bateson would have telexed this remarkable news to the International Astronomical Union in the United States, but, as it happened, no commercial telexes could be sent from New Zealand at night. (This was in the days before e-mail.) He did, however, get word immediately by telephone to astronomers in Australia. Meanwhile a Canadian observer—a professional astronomer, Ian Shelton of the University of Toronto, working at the Las Campanas station in Chile—happened to develop a photographic plate at 5:31 p.m. UT, some three and a half hours earlier. The photographic plate was of the Large Magellanic Cloud, and on it was an image of what appeared to be a very bright star. At first Shelton thought it was a flaw in the plate. He went outside to look and, sure enough, the new supernova was visible to the naked eye. Thus Shelton became the second person to discover SN-1987A, with Jones a close third.

The first person actually to see the supernova, an hour and a half earlier, was another Las Campanas observer named Oscar Duhelde, who did not tell anyone until Shelton brought the matter up. Because Jones did not see the supernova the night before, he was able to narrow down the instant of the explosion. Unknown to anyone at the time, remnants of the supernova first made their appearance in our vicinity at 7:35:41 a.m. UT on February 23, nearly two hours before Jones's nonobservation. These remnants were in the form of ghostly, elusive neutrinos.

At the Fairport mine in Painesville, Ohio, some twenty miles from Cleveland, miners and other employees of the Morton Salt Company ("When it rains it pours") often wear patches on their overalls and miner's helmets that read "Think Snow!" This is not because these people have a special affinity to skiing. Rather it is because they mine rock salt, which is sprinkled on icy roads—when it pours, they reign. The Fairport mine is very much a working commercial mine. Visiting it is a serious matter. The increasingly uneasy visitor is given a safety lecture that includes an interlude on using a rather heavy catalytic converter, worn on a thick leather belt at all times, when one is underground. In the case of a mine fire, the converter transforms carbon monoxide, which is poisonous, into carbon dioxide, which isn't. It also gets very hot when it is working and, one is told, despite that, one had better keep it in one's mouth in order to survive. One is then given a pair of high-top shoes with iron toes (in case one trips on some lumps of salt), a pair of blue coveralls, and a hard plastic helmet with a miner's lamp attached to it. One is then ready to go underground.

The mine shaft is two thousand feet deep—nearly two Empire State buildings down. There are two elevators that connect the bottom to the surface: a freight elevator—a sort of bucket—that brings up the salt and can be used in emergencies for people, and a passenger elevator that is big enough to hold ten miners on each of its decks (about forty miners at a time are at work in the mine) and, if necessary, various kinds of material. It is a comfortable if austere steel cage.

The ride down takes about four minutes and feels to the newcomer as if it takes an eternity. The temperature increases as one goes down. The walls of the working face of the mine are at a temperature of about eighty-five degrees Fahrenheit. There is a good deal of humidity at the bottom of the shaft. Murky tunnels seem to lead in all directions. One passes a large noisy fan and is led down a dark tunnel with only a miner's lamp illuminating the way. One has to be careful not to slip on the humid salt. After walking what seems like another small eternity, one is confronted by what looks like an impassable steel wall. There are warning signs about electrical equipment, but with a proper key the iron barrier can be opened and one can pass through. Once one does so, one is at first startled by the bright sodium vapor lights and the air-conditioning. One is suddenly in another world, the world of the high-energy physics laboratory.

The laboratory is devoted to a single enterprise: the care and feeding of an entity known as the "IMB detector." The initials IMB stand for Irvine, Michigan, and Brookhaven—that is, the University of California at Irvine, the University of Michigan at Ann Arbor, and the Brookhaven National Laboratory at Upton, Long Island, New York. These three institutions built the detector and maintain and run it. But what is it?

In the first place it is located in a separate chamber that one enters after passing the steel barrier. It was blasted out of the rest of the mine. The work was completed in 1980, seven years before the supernova explosion. The chamber was partitioned into four separate areas. First one passes through a large room filled with a very complex water filtration system, the reason for which will become evident shortly. In the next room is a typical laboratory bench with the obligatory microwave oven suitable for making coffee. Following this is a room that contains a maze of computer equipment, and finally one comes to the heart and soul of the detector: a gigantic pool of translucent and highly purified water. The tank is rectangular, 60 feet wide by 80 feet long, and holds a total of 8,000 metric tons of water. The water is so clear that when divers from the University of Michigan, who service the tank about every three weeks, first went into it, they got vertigo. It was like falling through thin air. This water is scanned by an array of 2,048 photo-multiplier tubes. These are very large hemispherical tubes, 8 inches in diameter, costing about a thousand dollars each, that were specially designed to detect very weak light signals. A few of them certainly cost more than Mr. Jones ever spent on all his homemade astronomical equipment put together.

What sort of light do these tubes detect? Therein lies a tale. Einstein's theory of relativity is predicated on the proposition that no material object can move faster than the speed of light in a vacuum: about 300 million meters a second. Note the word "vacuum." Light in a medium moves more slowly than light in a vacuum, which makes sense since its propagation is impeded. Thus it can happen that a radiation-emitting particle such as an electron can actually move faster than the speed of light in the medium in which the electron is moving. This sort of radiation is known as "Cerenkov radiation," named after the Russian scientist Pavel Alekseyevich Cerenkov, who shared the 1958 Nobel Prize in Physics for proposing it. It is visible, for example, as an eerie blue glow in pools that hold spent reactor fuel. The radioactive fuel elements emit electrons that move faster than the speed of light in water. The radiation is emitted in a narrow cone—something like a sonic boom—with its axis in the direction of the electron's motion, a detail that is crucial to using it in a detector. By detecting the Cerenkov radiation, it is possible to reconstruct the trajectory of the particle that emitted it, such as a rapidly moving electron.

The IMB detector had been designed in the first instance not to detect emissions from supernovas but to test the predictions of a then new class of speculative elementary particle theories. These theories were an ambitious attempt to unify various apparently disparate parts of elementary particle physics. It turned out that a seemingly inevitable consequence of this unification was that all matter had to be slightly unstable. The atomic nucleus is made up of neutrons and protons. Up to the time of these speculations—and to this day they remain speculations—it had been an ineluctable canon of elementary particle physics that the proton was absolutely stable. But these speculative theories predicted that the proton would actually decay, though incredibly slowly. The age of the universe is usually taken to be about 10^{10} years. According to these theories it would take a typical proton about 10^{31} years to decay. Not many protons in an average sample would decay in a year, which is just as well since we are made up of them. You would have to have a huge number of protons to see any effect. Enter the tank.

Most of these speculative theories predicted that if the proton did decay, one of the decay products would be a very rapidly moving positive electron—the electron's anti-particle with the same mass but the opposite charge to that of the electron. Now the plot becomes clear. The eight thousand tons of water in the tank contained about 2×10^{33} protons. If each proton, on average, took even as long as 10^{31} years to decay there would be enough protons in the tank so that one might expect about two hundred decays a year—an event every couple of days, sufficient to work with.

It may have occurred to the reader to ask why one should spend the $200,000 it cost to build such a tank and then locate it in a salt mine. (The physicists, incidentally, got a bargain, since Morton Salt wanted to try out a new piece of digging equipment.) Why not put it in some pleasant location on a college campus? The reason is what physicists call "background." The earth's atmosphere is being constantly bombarded by cosmic rays, and these can produce energetic positive electrons that can look just like the ones from proton decay. The 2,000 feet of material above the tank acts as a shield against cosmic rays. It is almost impenetrable. The "almost" is crucial to what happened on February 23, 1987 at 7:35:41 a.m. UT. We will come back to this shortly, but just to finish the story of proton decay. The tank was filled with ultra-pure water on July 30, 1982, and the experimenters began to wait for proton decay events. They are still waiting. Maurice Goldhaber, who was the senior experimenter from Brookhaven at the time, remarked, "We had some candidates, but they weren't elected." The experimenters concluded that if the proton decays by emitting a positron, as suggested by some of the unified theories, the experimental lifetime is in excess of 10^{32} years, which rules out many of the theoretical models.

Now back to February 23, 1987. As mentioned, the star Sanduleak -69° 202 was an ordinary-looking bluish object. It had a mass of about twenty solar masses and a radius about fifty times that of the sun. Astronomers are still puzzled as to why a blue star, as opposed to a red one, blew up. It is generally agreed that the sequence of events is as follows. Stars produce energy by a series of nuclear fusions. In these reactions nuclei combine to produce new nuclei with a release of energy. But for reasons involving the way in which protons and neutrons are bound in nuclei, this process stops at iron. Thus stars such as Sanduleak develop an iron core. But if this core becomes too massive—greater than ten solar masses—it cannot sustain its own weight and suddenly collapses, a process that takes less than a second. The core contains electrons as well as neutrons and protons. But because of the huge densities produced in the collapse, these electrons are forced to merge with the protons. This electron capture produces a neutron and a neutrino. To be more precise, it produces an anti-neutrino. This is important because if this anti-neutrino hits a proton, it can produce a neutron and an anti-electron. That is crucial to the observations we are about to describe.

The shock wave produced by the explosion of the core generates radiation. The supernova becomes visible—indeed, to the naked eye. But it takes time for this light to work itself out of the material—a day or so. The neutrinos—and anti-neutrinos—interact very weakly with matter. A neutrino can pass through many light years of lead without making a single interaction! It also moves at speeds at or close to the speed of light. Thus the neutrinos escape the supernova before the light does. They are produced in huge number in the explosion—some 10^{15} per square meter. If they did not interact so weakly, they might irradiate us. As it is, they pass right through us—and the earth—barely making a whisper. But they do make a whisper. At 7:35:41 a.m. UT on February 23—a day before the explosion was detected visually—8 anti-neutrinos which had been traveling for some 170,000 years since the explosion were detected in the tank. They all arrived within an interval of about six seconds. (Eleven additional anti-neutrinos were detected at the same time in a detector located in a zinc mine in Kamioka, Japan.) How were they detected? The anti-neutrinos interacted with the protons in the water in the tank. This produced positrons—anti-electrons. When such a positron finds an electron, the two annihilate and produce radiation quanta—gamma rays—of a known energy. These are detected by the photo-multiplier tubes on the side of the tank, and the whole event can be reconstructed. Because the detector had been running anyway, it found the anti-neutrinos. For the first time one was able to observe the inner details of the evolution of a star into first a supernova and then, once the protons had been transformed into neutrons, into a neutron star. It was a chance event. The last such nearby supernova explosion was some 400 years ago. It was witnessed by Kepler. Yes, it was a chance event, but, as Pasteur once said, in this sort of thing chance favors the prepared mind.

Einstein's Blunder

Our story begins in the fall of 1907. Einstein has become a technical expert "second class" in the Swiss National Patent Office in Bern. He had been promoted the previous year from "third class" and his annual salary raised to 4,500 Swiss francs. The year before that he had published the papers that had laid the foundations for twentieth-century physics. In the next half-dozen years he would move to the top of the academic ladder with an appointment in Berlin created especially for him with no formal teaching duties. But already he was well enough known among physicists so that in 1907 he had been asked to contribute a review article on relativity theory to the *Yearbook of Radioactivity and Electronics*. As he was beginning to work on this article he had what he later described as "the happiest thought of my life." He recalled many years afterward in a lecture that "I was sitting in a chair in the patent office in Bern when all of a sudden a thought occurred to me: 'If a person falls freely he will not feel his own weight.' This simple thought made a deep impression on me. It impelled me toward a theory of gravitation."

We can rephrase Einstein's happy thought in several ways which will enable us to see why it made such a "deep impression." Suppose we are on a roof with a group of friends and the roof suddenly collapses so that we all fall freely. (We always ignore effects of air resistance.) All of us will fall with the same acceleration with respect to the ground—the famous thirty-two feet per second per second. Thus none of us will be accelerated with respect to another. The only acceleration will be that of the ground accelerating toward us. If we have scales and weigh ourselves in midair, the scales will all read zero. We will be weightless. Gravity will have disappeared.

Here is another way of describing this—a "thought experiment"—that Einstein used and is known as the "Einstein elevator." We have a box in outer space. On the top of the box is a cord which can be pulled upward by a "being" (Einstein's term). If we switch on a suitable uniform gravitational field, all the objects will fall downward with the same acceleration. If we switch off the field, and have the "being" pull upward with this acceleration, the floor of the elevator will accelerate upward. If we are standing on a scale on the floor, the two situations will give us the same weight. This is what Einstein called the Principle of Equivalence. It was a new kind of relativity principle that related an accelerating frame of reference to a gravitational field.

If we accept this principle as fundamental, we have an explanation for an otherwise very puzzling fact. Ever since Galileo called attention to it, the observation that all objects, independent of their mass, fall with the same acceleration in a uniform gravitational field has been an accepted part of physics. Einstein seems to have been the first person to emphasize how strange this is. The reason is that two entirely different uses of mass are involved. On the one hand, "mass" is a measure of the inertia of an object. An object with a large inertial mass is more difficult to accelerate than one with a small one. But there is another use of mass here which is referred to as the "gravitational mass." The strength of the gravitational force acting on an object is proportional to this mass. The fact that all objects in a gravitational field fall with the same acceleration is only possible if these two types of masses are identical. If we accept the principle of equivalence as fundamental, these masses must be identical.

One of the great things about Einstein—at least until the last twenty years of his life—was his ability to propose experimental tests for his ideas. In his 1907 yearbook article, where the equivalence principle was first mentioned, he also proposed novel ways for it to be tested. These same test proposals were to recur in new guises over the next decade. We can best discuss them here using the Einstein elevator. The first one is the shifting of the frequency of light—red shift or blue shift—in the presence of gravity. I will not give Einstein's 1907 argument, which is a bit abstract, but rather how he might have argued, even then, and how we would present this to a class now. We would imagine a light quantum—the reality of which was the subject of one of Einstein's 1905 papers—in our elevator in space. Let it be generated at the top of the elevator and then detected at the bottom. If the "being" pulls up on the elevator with a constant acceleration, the detector will have gathered speed so that the light quantum will appear Doppler shifted—to the blue in this case; red if the "being" pushes the elevator down. But the principle of equivalence tells us this will also happen if we replace the "being" by a gravitational field. Einstein suggested that light escaping from the sun would be gravitationally red-shifted. It is very difficult to sort this effect out from others involving light escaping from stars, but it seems to have been observed.

He used the same ideas to predict that gravity bends light. Once again, imagine light coming in from the side of the elevator. If the elevator is accelerated upward, the floor will rise to meet the light. As observed by someone on the floor, the light will appear "bent." By the principle of equivalence, the equivalent gravitational field must then bend light. Einstein noted this in his 1907 paper and more or less left it at that. For the next four years he published nothing about gravity. Meanwhile he had first taken an associate professorship in Zurich, followed in 1911 by a move to Prague where he was appointed a full professor at the German University. That year he published his first detailed account of the principle of equivalence in a paper he called "On the Influence of Gravitation on the Propagation of Light." In this paper he presents a calculation of the bending of light by the gravitational field of the sun. What he did not know, until some Nazi-oriented German scientists brought it out later to discredit him, was that the same calculation had been done in 1801(!) by the German astronomer Johann Georg von Soldner. Soldner was answering Isaac Newton's query posed in his *Optics,* "Do not Bodies act upon Light at a distance, and by their action bend its Rays; and is this action *caeteris paribus* [all things being equal] strongest at the least distance?" Both Newton and Soldner had in mind a particle theory of light. Because of the equivalence of inertial and gravitational masses, the mass of the light particle cancels out and the answer, which is .87 seconds of arc,* depends only on the mass and radius of the sun. But in his 1911 paper Einstein proposed a test. He writes, "As the fixed stars in the parts of the sky near the Sun are visible during total eclipses of the Sun, this consequence of the theory may be compared with experience." In short, you would photograph a field of stars at night and then photograph the same field of stars with the sun in front of it but eclipsed by the moon. What you should find, Einstein claimed, was a tiny shift in the position of these stars—outwards—because the light is bent toward the sun.

The year after he published this paper, Einstein was appointed a full professor at his alma mater, the Swiss Federal Institute of Technology, the ETH, in Zurich. But by 1914 he had moved to Berlin, where he would stay until in 1933 he came to

*Einstein made a slight numerical mistake and found .83, while Soldner found .84.

the United States. One of his new colleagues in Berlin was a young astronomer named Erwin Freundlich. Freundlich had begun corresponding with Einstein about astronomical tests even before Einstein came to Berlin. But now the two men began planning an expedition for Freundlich to the Crimea, where there would be a total eclipse on August 21, 1914. Einstein helped him to get financing, some of which came from the German arms manufacturer Krupp. In July 1914 the expedition left for Russia, and in August World War I broke out. Freundlich and his team were arrested by the Russians and held as prisoners of war. The eclipse measurement was never carried out. If it had, Freundlich would have found an effect, but *twice* what Einstein's Newtonian theory predicted! Chances are that Einstein would not have been overly surprised because by this time he had made serious progress toward his scientific masterpiece, the general theory of relativity and gravitation, toward which the 1911 paper was only a stepping-stone.

I will not try to review all the starts and stops that led to the creation of this theory in any detail. But I would like to give you a sense of it so you will be able to appreciate the "blunder." Einstein understood from the beginning that the principle of equivalence could not be the whole story. It applied to only one type of gravitational field—a uniform one. No fields in nature are truly uniform. Indeed, what Einstein did in his 1911 paper was to *assume*—he says this explicitly—that the formulae he derived for the uniform field could be applied to those cases, such as the bending of starlight by the sun, where the field is not uniform. But, aside from this, he began to get the intuition that geometry—in this case the geometry of space-time—would play an essential role. We can see this already from the Einstein elevator. Again, imagine light coming in from the side while the elevator is accelerating upward. To an observer in the elevator the light ray will appear bent, hence it will also be bent in the equivalent gravitational field. But the paths of light rays traveling in a vacuum are what we mean by straight lines. In the absence of gravity, if we constructed a triangle made of three light rays, it would obey the laws of Euclidean geometry. In particular, the interior angles would add up to 180 degrees and, for a right triangle, the Pythagorean theorem would hold. But if we construct a triangle of light rays in a gravitational field, it will not obey the laws of Euclidean geometry. The interior angles might be anything, depending on how the light rays are curved. Indeed, as early as 1900 the German astronomer Karl Schwarzschild, who was later to make very important contributions to Einstein's theory, proposed testing the applicability of Euclidean geometry to actual space by studying the angles of a triangle composed of a nearby star and two points on the earth's solar orbit. He was unable to draw any definitive conclusions from his data.

Einstein had another argument pointing to a changed geometry. He imagined a wheel that was sent spinning at a very high speed. Relativity predicts that the rim will contract while the diameter will not. Thus the ratio of the circumference to the diameter will no longer be pi—the Euclidean answer. But we can replace this by a suitable gravitational field and again conclude that gravity "warps" space. Thus, not long after he published his 1911 paper, Einstein decided that he would need to learn more mathematics—the possible geometries of four-dimensional spaces. Here, fortunately, he was able to get help.

Marcel Grossmann had been a fellow student of Einstein's at the ETH. Indeed, he was the student who took the careful notes that Einstein borrowed when he needed to pass examinations. Grossmann became a mathematician and by 1907 had become a full professor of geometry at the ETH. By 1911 he had been appointed dean of the mathematics and physics faculty. One of the first things he did was to recruit his classmate to return to the ETH, which Einstein did for three semesters. By the time Einstein returned, he had a general feeling for what he might need for a new theory of gravitation, but not the details. He said to Grossmann, "You must help me, or else I'll go crazy." As was often the case, Einstein was unfamiliar with the literature. In this case, that pertaining to these geometries, he had, as a student, taken a course in the differential geometry of surfaces embedded in ordinary three-dimensional space. He understood that the key to characterizing the curvature of these surfaces—think of the surface of a globe like that of the earth—was the expression that gave the distance between two very closely separated points: the generalization of the Pythagorean theorem. For the Pythagorean—the Euclidean—case we simply have the sum of the squares of the coordinates. For the general surface, the squares of the coordinates are multiplied by expressions that characterize the geometry. If the surface is not smooth, you can also have "cross terms" that link coordinates relating to different directions. All of this can be summarized in a single mathematical expression called the "metric tensor." This tensor determines everything about the geometry. Einstein, it seemed, knew some of this before he spoke to Grossmann. He also knew that somehow gravity must determine this tensor. But how? Grossmann told him to read the works of the nineteenth-century German mathematician Bernhard Riemann and his successors. Riemann, who died in 1866 at the age of forty, was one of the greatest mathematicians who ever lived. In 1854 he gave an inaugural lecture at Göttingen in which he presented elements of the theory of these higher-dimensional curved spaces. His successors developed these ideas. There was a whole literature that Grossmann reviewed in a joint paper that he and Einstein published in 1913. These concepts provided the vocabulary, but it took Einstein another two years to put them together into a coherent structure.

There were various reasons for the delay. The mathematics was new and, for the time, very difficult. Furthermore, in these general four-dimensional geometries space and time could mix together, like the cross terms in different directions for the nonuniform surfaces. This made the theory hard to interpret except in regions where the gravitation was approximately uniform and space and time could be separately identified. There were also conservation laws, such as the conservation of energy, that the theory had to preserve. In addition, Einstein finally decided that all the equations should have their form preserved—something that is called "covariance"—when the space-time coordinates are transformed in an entirely arbitrary manner. This limited the mathematical expressions allowed into the theory. Indeed, by 1915, Einstein realized that his set of conditions almost completely determined the possible equations of the theory. The remaining ambiguity, which he put aside for the time being, is what led to the "blunder," as we will soon see.

The Einstein equations determine the geometry of space-time from the distribution of gravitating matter, but they do not explain how anything moves. To deal with this Einstein made a further assumption, which he was able to derive from first principles only in the late 1930s. He assumed that light, and other objects, followed the straightest curve allowed by the geometry—what mathematicians call a "geodesic." In Euclidean geometry, where the space is not warped, this is the shortest distance between two points. On the surface of a sphere it is the arc of a great circle, such as the equator. In the four-dimensional space-time of Einstein's theory it is the *longest* "distance" between the coordinates in space and time that describe the places and times of two separated events. If you happen to have a clock with you when you travel between two points on the geodesic, it will record a longer time than the time interval along any neighboring path. Note that in this theory there are no forces. Motion is determined by the geometry. This became the model for Einstein's later work on unifying fields, and the motif is with us still.

Once in possession of these equations, Einstein was now in a position to begin to extract the consequences—a process that continues to the present day. The first thing he noted was that the Newtonian laws of gravitation emerged as a first approximation, an absolute necessity since that theory had worked well for nearly three centuries. But new physics emerged with the next approximation. In the first place, Newton's law of gravity was modified in such a way that planets no longer followed closed orbits like the elliptical orbits first discovered by Kepler. Actually the closed ellipses ignored the effects of other planets. Taking these into account, the ellipses slowly precessed. If the orbit of a planet like Mercury—where the effect was largest because of its proximity to the sun—was plotted over eons, the curve would look something like the petals of a flower. But there was an anomaly. These known planetary effects did not account for the entire observed precession. There was a small residual amount which astronomers could not account for. Various suggestions, such as introducing an unobserved planet, had been proposed to explain it. One can only imagine Einstein's joy when his theory gave the exact amount of the anomaly with no additional assumptions. He also used his equations to compute the deflection of starlight by the sun and found an answer that was *twice* the result he had published in 1911. This was verified by two eclipse expeditions in 1919, which made Einstein world-famous. But two years earlier he had published a paper entitled "Cosmological Considerations on the General Theory of Relativity," which began the era of modern cosmology and in which he made his "blunder."

Cosmology is the study of the universe at large, its origins and its destiny. In 1917 most astronomers believed that the material universe consisted of only our galaxy, the Milky Way. Whatever was outside this galaxy was taken to be empty space. There were exceptions to this consensus. In 1914 the American astronomer V. M. Slipher announced that he had found that some spiral nebulae were receding so rapidly that it seemed unlikely they could be contained within our galaxy. But it was only in the next decade when the observations of the great American astronomer Edwin Hubble persuaded most people that there really were "island universes" outside the Milky Way. The problem that Einstein faced in 1917 was why such a Milky Way universe was stable. This is a question that goes back at least as far as Newton. It is related to the fact that all objects *attract* each other gravitationally. Even in a universe where, on the *average,* there is no motion, there will be motions proper to individual stars which average out. This means that there will be places where matter will temporarily cluster. But these material clusters will attract more matter gravitationally, and over the course of time the universe will collapse into one or maybe a few such gravitational clusters. To prevent this, Newton proposed to make his universe spatially infinite. "But if the matter was evenly disposed throughout an infinite space," he wrote in 1962 in a letter to Richard Bentley, ". . . some of it would convene into one mass and some into another. . . . And thus might the sun and the fixed stars be formed." Of course he believed that the universe was finite in time, since it had a beginning with the Creation.

By the late nineteenth century doubt was cast on Newton's resolution, and it was even proposed that the law of gravity be modified to add a term that would supply a force to counteract gravity. Einstein faced this same dilemma with his theory of gravitation, but to it he added another, which to us seems quite strange. He noted that there would be stars that would be moving rapidly enough to escape the gravitational hold of the galaxy and wander off to infinity—a kind of evaporation of the galaxy! To resolve these perceived difficulties he proposed, in his 1917 paper, to modify his gravitational equations. Here came the "blunder."

Einstein wanted to modify the equations but preserve all the basic properties—the symmetries—of the original ones. It turns out that this limits the possibilities to one single new term. This is a term that is proportional to the metric tensor itself—something that Einstein called the "cosmologic member"*—with a dimensional constant of proportionality that we now call the "cosmological constant." As far as he knew, and as far as we

*The German is *kosmologische Gleid,* where one might translate *gleid* as *term.* But in Einstein's book *The Meaning of Relativity,* this is how it is translated.

know, there is no a priori way of fixing this constant. Einstein was free to choose it both in magnitude and sign as he saw fit.

To this end he made another assumption, one which is basic to all modern cosmologies. He assumed that the average distribution of matter in the universe was both homogeneous and isotropic. Put in simple terms, he assumed that any observer would see the same distribution from any location in the universe, and that this distribution would look the same in every direction. Einstein would have had a hard time justifying this in detail. But by now we have observed a vast amount more of the universe than was available to him, and the more of it that is studied, the more correct does this assumption appear. It vastly simplifies the equations. This uniform density of matter interacts with itself gravitationally, and fluctuations from the average are what would produce the unwanted instability. So Einstein chose the cosmological constant to have just the magnitude and sign needed to counterbalance the gravity.* The cosmological "member" is a kind of anti-gravity and produces a pressure that in Einstein's model just counteracts the gravitational pressure. This fixes the magnitude of the constant in terms of the mass density of the universe. With this choice Einstein's universe is, apart from small proper motions of the stars, stationary.

Even during World War I, Einstein was able to carry on some scientific correspondence with colleagues outside Germany. Some of this had to do with his new cosmology. Two of the most significant cosmological correspondents were the mathematician Hermann Weyl in Switzerland and the astronomer Willem de Sit-

*The relation is $\lambda = \rho \times 8\pi G$ where λ is the cosmological constant and G is Newton's gravitational constant which equals $6.672 \times 10^{-11} m^3 kg^{-1} sec^{-2}$ and ρ is the average matter density of the universe. Thus λ has the dimensions of inverse seconds squared. If we put in for the density the approximate value $2 \times 10^{-28} kg\ m^{-3}$, we see that the constant Einstein needed was about 10^{-38} inverse seconds squared. One often finds this constant expressed in units of inverse lengths by dividing it by the square of the speed of light. It is entirely unclear what one is to make of such a dimensional numerical constant.

ter in Holland. Soon after de Sitter read Einstein's paper, he realized that there was a stationary cosmology that Einstein had missed. On its face this is a very strange cosmology—Einstein certainly thought so—but in a new, nonstationary guise it is still with us. In the de Sitter world the average matter density is taken to be zero but the cosmological constant is not. You would think offhand that this world would be unstable and would expand since the cosmological term would supply a negative pressure with nothing to counterbalance it. In general you would, as we shall see, be right. But you can choose the curvature of this world in just such a way that this negative pressure is balanced, which is what de Sitter did. From our point of view, this is a pretty unnatural choice. But de Sitter made a remarkable discovery, which he published in 1917 in a rather obscure appendix to one of his papers. While the de Sitter universe as a whole may be stationary, if you introduce a test particle into it, the particle will be accelerated. This is not as crazy as it sounds since the particle will be subject to the negative pressure of the cosmological term. Indeed, this brings up an important point: the distinction between the behavior of space as a whole and of the material objects within it. As Einstein used to say, "Space is not a thing." The universe can expand at speeds greater than the speed of light, though no material object within it can exceed this speed. De Sitter noted that this acceleration would also happen for light from "distant stars or nebulae," which would appear red-shifted. This seems to be the first reference to the cosmological red-shift of light that would henceforth dominate cosmology. Before Weyl made his contribution, a completely novel idea was introduced by a totally unknown scientist—not even a physicist—in St. Petersburg (which in 1924 became Leningrad).

Alexander Aleksandrovich Friedmann was born in 1888 to a talented musical family in St. Petersburg. He took his degree in mathematics from the university in 1910 and then went into theoretical meteorology, publishing in 1914 a seminal paper on temperature inversion in the upper atmosphere. When the war broke out, Friedmann volunteered for an aviation detachment and flew in military flights as an observer. By 1917 he became section chief and then director of the first Russian factory that manufactured aviation instruments. In 1920 he returned to St. Petersburg where eventually he became director of the Physics Observatory. He died in 1925, at the age of thirty-seven, from typhoid fever. George Gamow, who had signed up to take a course in relativity from Friedmann the year he died, claimed that Friedmann had taken a flight in a weather balloon where he took sick and did not recover.

After Einstein's theory was confirmed by the eclipse experiments, Friedmann decided to teach himself relativity. In 1921 a text on relativity had been published by the German Nobelist Max von Laue, which included the general theory. Laue had even come to Switzerland a decade earlier to visit Einstein in the patent office; in the 1930s he was one of the few German physicists who had the courage to resist the Nazis. He taught relativity throughout this period. We know that Friedmann read Laue's text because he cites it in his 1922 paper, the one that produced a "Copernican revolution" in cosmology.

Unlike Einstein and de Sitter, Friedmann had no a priori prejudice that the universe had to be stationary. He made the same assumption that Einstein did about the homogeneity and isotropy of the matter density in his world. But he allowed this density to vary as a function of time.* He did not specify how this density varies in time, something for which we need the Big

*There is an interesting conceptual question about what homogeneity and isotropy—which incidentally implies homogeneity—mean in a universe that is not stationary. As we peer into this universe from any vantage point, we see backward in time to various epochs that do not resemble our own. What one presumably means is that any observer who made a record of what he saw of the Big Bang from a given vantage point would produce the same record as any other observer, and that at a given time these records would be the same in all directions. The Friedmann papers in translation can be found in *Cosmological Constants,* edited by Jeremy Bernstein and Gerald Feinberg (New York, 1986).

Bang model. But he derived the general equations for the function of time that measures the scale of the universe in terms of this density. He noted that Einstein's and de Sitter's models were just special cases of his equations. In his 1922 paper he discussed two cases. The first case he called the "monotone world," which begins in a point and expands indefinitely, and a "periodic world," in which the expansion stops after a finite time. He noted that "Our knowledge is insufficient for a numerical comparison to decide which world is ours." Many cosmologists would agree that this is still true, though the data are getting better.

Friedmann sent his paper to Einstein, whose reaction was quite remarkable. Einstein decided that Friedmann had made a mathematical mistake and that there was no expanding universe in this model. He even published this in a short note in September 1922. He soon got a letter from Friedmann that sorted him out, and he published a retraction in May 1923 in which he said, "I am convinced that Mr. Friedmann's results are both correct and clarifying." By this time Hermann Weyl had noted that in de Sitter's model not only was there a cosmological red-shift but, at least for nebulae that were not too far away, the magnitude of this shift was proportional to the distance from the observer. This is usually referred to as "Hubble's law" in honor of the fact that Hubble confirmed it in his great observational paper of 1929. Hubble himself referred to it as the "de Sitter effect." Considering the scatter of his data, one wonders whether he would have been so confident of this linear relation without de Sitter's theory.

Einstein's first reaction to his cosmological constant following his episode with Friedmann is, as far as I know, in a 1923 letter to Weyl. In it he wrote, "If there is no quasi-static world, then away with the cosmological term!" In 1924 Friedmann published a second paper in which he considered those cases he had not treated in his first paper, thus completing the entire spectrum. George Gamow reports in his autobiography that he had conversations with Einstein several years later in which Einstein referred to the introduction of the cosmological constant as his biggest scientific "blunder." The last published word we have from Einstein himself was in the second edition of his book *The Meaning of Relativity,* which appeared in 1945 and to which Einstein had added an appendix on cosmology. In the appendix he presents a wonderfully lucid account of Friedmann's work. There is a footnote that reads, "If Hubble's expansion had been discovered at the time of the creation of the general theory of relativity, the cosmological member would never have been introduced. It seems now so much less justified to introduce such a member into the field equations, since its introduction loses its sole original justification—that of leading to a natural solution of the cosmologic problem." That would seem to settle the matter. But did it?

If one were writing a novel about this, rather than a history, one would hesitate to include this next chapter since one would worry that no one would believe it. To put the matter simply, Einstein's "blunder" is now the hottest topic in present-day cosmology. It has raised puzzles so deep that their solution may take us into entirely new realms of physics. How did this happen?

To understand it, we must say a few words about the quantum mechanical "vacuum." Apart from its usefulness for making jokes about getting something from nothing, and the like, the term "vacuum" is in this context rather misleading. The "state of lowest energy" might be better. A simple example of what is involved is the quantum theory of the harmonic oscillator—as motion that oscillates back and forth at some fixed frequency. The classical harmonic oscillator has a lowest energy state in which the energy is zero—a real vacuum. But the quantum mechanical oscillator always has a residual energy, which is usually called the zero-point energy. A collection of such oscillators can never be at the absolute zero of temperature because of this residual energy. When quantum theorists developed a quantum theory of fields they found the same phenomenon, which is not surprising since one can analyze these fields in terms of collections of oscillators.

The state of lowest energy here—which is what physicists call the vacuum—is a very dynamic place with virtual particles being created and destroyed in it and producing an energy that has measurable effects. If a quantum field has a nonvanishing vacuum energy, this will show up in general relativity precisely as a cosmological constant. Thus if quantum mechanics is right, you will have a cosmological constant whether you want one or not.

The question is whether it is big enough to have any observable effects. Most cosmologists believe that there was one epoch very close to the Big Bang where this effective cosmological constant was so large that it dominated the expansion completely. The best guess is that this took place at something like 10^{-35} seconds after the Big Bang. When this happened the universe became an expanding de Sitter space since the matter density in it was negligible. In fact this expansion, during the very brief time it lasted, was exponential—huge. This epoch—which cosmologists call "inflation"—enables us to account for some of the observed regularities in the cosmic background radiation—the famous three-degree radiation left over from the Big Bang. This use of the cosmological constant has been around now for some twenty years and has become almost "classical," though as yet there is no single compelling model. What is new is very new. It is so new that one should have a bit of caution about it. It revolves around the effects of a cosmological constant today.

Here we must divide the discussion in two. On the one hand there is the theoretical question of why at the present epoch we should have a cosmological constant of such a magnitude as to compete with the gravitational effects of ordinary matter. On this point I think it is fair to say that no one has the slightest idea. It does not seem to be compatible with our general ideas about the quantum theory of fields. On the other hand there is the question of whether or not observations tell us that we need such a cosmological constant. Here at least we can state with some certainty what effects to look for. The effects are all in one way or another related to the fact that the negative pressure associated with a cosmological constant causes the present universe to accelerate. This implies several things. Here are three: The Hubble curve would no longer be given by a straight line. The lifetime of the universe would be different from the one predicted by the theory without the cosmological constant—indeed, longer. Finally, distant objects, which were created closer to the beginning of the universe, would be expanding more slowly than nearer objects, which are being accelerated. Remarkably, there are serious claims that all three of these effects have been observed.* These are very difficult observations, but if they hold up, Einstein's frog will have turned into a prince.

*The literature on this is now so vast that to keep up with it one has to make use of the resources of the World Wide Web. If you do so, you will find articles at all levels.

ON SCIENTISTS

Over a span of many years I have had the opportunity of knowing, or at least meeting, several geniuses. There was Wim Klein, the Dutch-born computer programmer who worked at CERN, the elementary physics laboratory near Geneva. Mr. Klein did extraordinary feats of mental arithmetic. You could ask him the fifteenth root of a twelve-digit number and he would rapidly compute it to high accuracy in his head. This required, among other things, knowing the table of logarithms by heart. There was Hans Eberstark, who also lived in Geneva and earned his living as a simultaneous interpreter. He knew twenty-five or thirty languages and was at least as good as Klein in doing mental arithmetic. There was Duke Ellington, whom I knew as a young teenager, and Jacques Brel, whom I came to know much later. These men could sit down with their respective musical instruments—piano and guitar—and improvise music that would be played or sung for generations. There was W. H. Auden, whom I met briefly when I was at the Institute for Advanced Study, who used language in ways no one else could have thought of. There is the artist Helen Frankenthaler and the chess player Sammy Reshevsky. Then there are the physicists. About them I can only think of an interview Isaac Stern once gave. He was asked how he would rank himself among all violinists. "Among all violinists," he replied, "I am second." "But who then is first?" "I cannot tell you," he answered, "because they are all my friends."

I have often reflected on what these people have in common. Two things stand out. The first is that they are all capable of doing things with apparent ease that most of us cannot do at all. This is something that cannot be taught. It is hard-wired. Mr. Klein, who was the most generous of men, often tried to teach me how to do mental arithmetic. He said it was merely a matter of knowing a few "tricks." But these tricks turned out to be numerical identities that arose spontaneously in Klein's mind as he did one of these calculations. Finding these identities was as hard as doing the calculation itself—except to Klein. The second thing is context. I can imagine a Sherpa who was born with the same mental ability as Einstein. But I cannot imagine such a Sherpa—at least in the days before the Internet—discovering the theory of relativity. He would not even know what questions to ask. The questions Einstein asked grew out of centuries of scientific discovery. Likewise, we must have a cultural context to know that Eberstark is speaking a "language" and is not some madman speaking in tongues. All the offerings in this section deal with people of genius. But they also deal in one way or another with the cultural context that enabled these people to flourish.

Enough Einstein?

Edith Sitwell is supposed to have said that there were only two things worth writing about, love and the eighteenth century, and that too much had been written about love. To this list I would adjoin "Einstein" and ask if too much has been written about him. This question has been inspired by three books I read in succession, which join a list now extending into the hundreds. Before discussing these books specifically I would like to give an extremely brief, very selective, and entirely subjective overview of what has preceded them. To make a full study would require yet another Einstein book. I will begin with the first Einstein biography I ever read and, indeed, one of the best I have ever read: *Einstein: His Life and Times* by Philipp Frank.

Frank, who was my first great teacher in physics, published his book in 1947, just before I took my first course with him. He had been born in Vienna in 1884, making him five years younger than Einstein. In 1907, the year he took his Ph.D., Frank wrote a philosophical paper that attracted Einstein's critical attention.

A correspondence began, and then a friendship, which finally led in 1912 to Frank's being appointed as Einstein's successor at the German University in Prague. Frank always liked to tell the story of how Einstein had given him the military uniform he was required to wear at his inaugural ceremony. Einstein had worn it once. Frank also wore it once. His wife gave it away during the war to a freezing army officer. Some years later Frank became a founding member of the Vienna Circle—philosophers and scientists who had joined forces to get rid of what they thought was metaphysics in science. He emigrated to the United States in 1938 and was appointed to a lectureship at Harvard. He died in 1966 at the age of eighty-two.

When he wrote his book, Frank had studied Einstein's physics for nearly half a century and understood it as deeply as anyone could. He was not really an historian, so his biography, which has no footnotes, is somewhat hazy on the finer historical details. It has also to be understood that Einstein was still alive when the book was being written and was well aware that Frank was writing it. While it is by no means a hagiography, if Frank knew anything about the darker sides of Einstein's life, he kept it to himself. But anyone who wants an introduction to Einstein biography should begin with this book.

While Frank was finishing his book, another Einstein project was unfolding. Paul Arthur Schilpp, a professor of philosophy at Northwestern University, had created a series called the Library of Living Philosophers. What he did was to select a group of noted philosophers, such as Bertrand Russell and John Dewey, who were still alive, and have people write essays about them with introductory and concluding essays by the subject. From what I have heard, Professor Schilpp must have been one of the most persistent people who ever lived. Not only did he have to corral the essayists, but in the case of Einstein had practically to chain him to his desk to write his introductory essay. There is no way one can express one's gratitude to Schilpp for this. The resulting essay, which opens the Einstein volume published in

1949, is called "Autobiographical Notes." It is one of the classics in scientific writing. Schilpp also had the brilliant idea of having the original German and the English translation side by side on facing pages so that one could get a sense of Einstein's style in the language he was most comfortable with. He almost never wrote in English. Nonetheless it is not really an autobiography. You will find nothing about either of Einstein's wives, nor of his children. You will find nothing about his career, nor even of his having been forced to leave Germany. The "merely personal," as Einstein called it, is almost totally absent. What you will find is a treatise on the development of modern physics by the man who was responsible for much of it, and how these ideas came about. But even here one is tantalized. Einstein tells us that he hit upon the essential paradoxes that led to the theory of relativity at the age of sixteen. But he did not discover the theory until a decade later. If only I had had a chance to talk to Einstein, at least after I knew something, I would have asked, "What kept you so long?"—what exactly was the triggering idea? We now know a little more about this, but not from his autobiographical notes.

Einstein died in 1955. One might think that this event would have been followed by a flood of biographies. It wasn't, really—at least not right away. The first ones appeared in the early 1970s. I am in a position to give at least a partial explanation for the delay since I wrote one of them. In my own case, I was approached by Frank Kermode, who was editing the Modern Masters series of relatively short biographical studies in various fields. My "master" was Einstein. The late British journalist Ronald W. Clark also completed his biography, *Einstein: The Life and Times* at about the same time. Both were published in 1973, and both of us ran into the same problem: Otto Nathan and the Einstein estate.

On March 18, 1950, Einstein drew up a will which left all his literary properties to two trustees, the late Otto Nathan and the late Helen Dukas. Miss Dukas had been Einstein's secretary since 1928 and had come to the United States with the Einsteins in 1933. From 1935 on she lived in the Einstein house in Princeton, at 112 Mercer Street, with his wife and stepdaughter Margot, and after Einstein's death she shared the house with Margot. Einstein's wife had died in 1936. Miss Dukas, whom I had gotten to know pretty well when I was at the Institute for Advanced Study in the late 1950s, was everyone's idea of the ideal maiden aunt. She was immensely kind and helpful to those of us who were trying to write about Einstein. I still cherish a long critical letter she wrote about my manuscript, correcting all sorts of errors—" 'his summer-home' [on Long Island] implies that he owned it—they were all rented." She was a woman of great dignity and a great sense of propriety. I make these points for a reason that will emerge. Otto Nathan was a different story.

Nathan was an economist, also a German refugee, who somehow had befriended Einstein. I do not know, but I would guess that Einstein made him a trustee of his estate because he wanted someone to look after the financial interests of his stepdaughter and Miss Dukas. I cannot imagine that Einstein had in mind the role that Nathan invented for himself, the guardian of Einstein's reputation. I got some hint of what this meant when in one of her letters to me Miss Dukas suggested that I show what I was writing to "my neighbor" (Nathan lived a block away), implying that I should get his approval. I had no intention of getting Otto Nathan's approval for anything, so I simply went blithely on my way. It should be mentioned that I was writing the book also as a profile for the *New Yorker,* and when it came time to publish in the magazine, its lawyer, Milton Greenstein, began negotiating with Nathan for permission to quote material owned by the estate. It should also be mentioned that what I wanted to quote was all in the published literature. I had no access to any private Einstein correspondence. I was simply using such material as quotations from his "Autobiographical Notes," which did indeed belong to the estate. Very soon Greenstein reported that there was trouble. As far as he could tell, Nathan seemed to want editorial control over what I wrote, which neither I nor the *New Yorker* had any intention of giving him. Greenstein advised me to speak to Nathan myself. I did, several times, which was like putting your head into a buzz saw. On many occasions Nathan told me that if I didn't like it I could simply paraphrase Einstein, which would have ruined everything. On one manic occasion he assured me that the estate had the rights to the formula $E=mc^2$. After nearly a year of this, sometime during the course of which Greenstein had suggested that I *not* talk any more to Nathan, it finally came down to money. Nathan charged the *New Yorker* the largest sum the magazine had ever paid for permissions. Nathan assured me that the *New Yorker* could afford it. I told him—which was the truth—that I was the one who would be paying for it, and I could *not* afford it. This made no dent. Nonetheless I consider myself lucky. Clark ran into the same problem, but he was not as lucky.

Clark was an enterprising journalist with no particular qualifications to write about Einstein. Judging from his book, he did not really have much understanding of the physics. But then neither did Dr. Nathan. The real trouble for Clark was that, according to Nathan's standards, he was not sufficiently respectful of Einstein. He had decided that Einstein's second marriage, in 1919 to his cousin Elsa Einstein Löwenthal, was troubled. His first marriage to Mileva Marić had been disastrous and had ended in an unpleasant divorce. Why Clark had come to this conclusion, which has some truth to it, I am not sure. He had very likely read the letter Einstein wrote in March 1955 to the sister of his lifelong friend Michele Besso. Besso had just died. Einstein was to die only a few weeks later. He writes, "But what I admired most about Michele was the fact that he was able to live so many years with one woman, not only in peace, but also in constant unity, something I have lamentably failed at twice." In any event, Clark made this into something of a melodrama, ending his book with, "And anyway, as Elsa had felt nearly twenty years before, dear God, it was too late now." Because of this and other perceived transgressions, Nathan decided to refuse Clark permission to quote from any material to which the estate had rights. Clark went to court—and lost. He had to rewrite his book.

Not content with dealing with the likes of us, Nathan also got into a battle with Princeton University Press. For twenty years the press has been engaged in the monumental task of publishing Einstein's entire written output—correspondence as well as other writings, published and unpublished. So far eight volumes have appeared, bringing the work up to the year 1918. Eventually there may be at least twenty-five volumes. Since each volume has taken something like two years to complete, one should not hold one's breath.* In 1971 the press signed an agreement with the Hebrew University in Jerusalem where, according to his will, Einstein's papers were to be permanently deposited. Ever mindful of his self-appointed role as keeper of the flame, Nathan sued to keep the press from publishing without his imprimatur. He was dissatisfied with the editorial arrangements, which he apparently believed gave too much authority to physicists and historians of science and not enough to himself. (He once told me that the project had too few editors and that Einstein needed at least as many as any of the other philosophers.) He dragged the matter through various courts until finally he lost in 1981, having delayed the project by several years. I will never forget the phone call I received from him on the weekend after my first article appeared in the *New Yorker*. He complained that he had not been thanked personally in the article. I reminded him that he had been informed by the *New Yorker* that this kind of acknowledgment did not appear in their articles. I then asked him, that aside, what had he thought of the article. "Oh," he told me, "I never read it."

Fortunately there were some historians of science who were

*The actual work, which includes collecting material, writing commentaries, and preparing camera-ready copy, is under the supervision of Professor Robert Schulmann of Boston University at the Einstein Papers Project in Boston. I am grateful to Professor Schulmann and also to Professor Gerald Holton for a careful reading of this essay.

willing to put up with Nathan in order to use the material that became available after Einstein's death. Three come immediately to mind: Gerald Holton, Martin Klein, and Abraham Pais. Holton, a professor at Harvard was, like myself, a disciple of Philipp Frank. Indeed, he helped to prepare the scientific diagrams in Frank's book. As far as I know, Holton was the first person who tried to bring order into Einstein's *nachlass,* which was casually stored in Princeton. One of the ironies in the Nathan story is that Einstein himself seemed rather indifferent to what happened to his papers. Holton used the material he found to write several wonderful essays that describe Einstein's approach to doing science. What emerges from them is Einstein's selective disregard of experiments that seemed to contradict his theories but which his instinct told him were probably wrong. During his period of greatest creativity, Einstein seemed to have a pipeline to the "Old One," his affectionate reference to God. I am not aware of any case in which, at this time, his intuition let him down. Physics differs from pure mathematics precisely because we do have to pay attention to experiment. Not many physicists can get away with trusting their intuitions even when experiments appear to contradict them. But then there was only one Einstein.

Martin Klein, a professor at Yale, taught us a great deal about Einstein's contributions to the creation of the quantum theory. It was well known that one of Einstein's great papers in 1905—the year he created the theory of relativity—had to do with the light quantum. This had been introduced by the noted German physicist Max Planck as a mathematical artifact to help him derive some properties of electromagnetic radiation trapped in a cavity. But Einstein argued that these quanta were real. In certain circumstances light behaved as if it consisted of elementary particles. It was for the work, and not for relativity, that Einstein received the Nobel Prize in 1921. Einstein understood that this meant the end of classical physics. He always said that the theory of relativity was not a radical departure from pre-twentieth-century physics. The departure came with the quantum.

In 1923, when Einstein received Louis de Broglie's paper suggesting that material particles like electrons also had a wave nature, he immediately realized the importance of this idea and how it could be tested. It is the principle behind the electron microscope. There was now a complete wave-particle duality, and Einstein spent the next several years thinking what this might mean. It was during this period that a new generation of physicists, such as Werner Heisenberg, Paul Dirac, and the somewhat older Erwin Schrödinger, created the modern quantum theory. Einstein, as Klein makes clear, was in some sense the muse for this development. He may even have independently discovered what became known as the Schrödinger equation which governs the propagation of these waves in space and time. He certainly never published it, but there is a wonderful exchange of letters between Einstein and Schrödinger. In one of them Einstein writes down what he thought the Schrödinger equation was and he criticizes it, proposing an amended version. Schrödinger was overjoyed because the amended version was in fact his original equation. Once the Schrödinger waves received their interpretation as waves of probability, Einstein began having serious doubts about the theory. In the end, it was not the probability aspect that bothered him but the fact that the theory refused to answer questions which Einstein thought any basic theory should answer. For example, why and when does this particular atom—not the average atom—spontaneously decay? He engaged in a series of colloquies with Niels Bohr that lasted decades. They are wonderfully described in one of the articles in Schilpp's collection.

In 1982 the late Abraham Pais published his monumental biography of Einstein, *Subtle Is the Lord: The Science and the Life of Albert Einstein.*"* The title comes from Einstein's famous apho-

*New York, 1982.

rism, originally remarked in German, "Subtle is the Lord, but malicious He is not," a reference to the fact that Einstein believed the universe was ultimately open to human comprehension. The subtitle is a reference to Pais's priorities in writing his book. The science comes first.

In writing his book Pais had every advantage. He had a profound understanding of the physics. He had the scholarly temperament required to read all the relevant documents in their original languages. He was a European—Dutch—so he shared with Einstein the feeling of what it meant to be a refugee, and he also shared Einstein's partiality for Jewish jokes. He was for several years Einstein's colleague at the Institute for Advanced Study and had many conversations with him. He does not seem to have asked Einstein much about the history, which is a pity since it would be nice if we knew more about it firsthand.

The book really cannot be read in full without a considerable knowledge of physics. The nonphysicist can certainly find in it a full-scale biography, but I would imagine that it must engender the same sense of frustration that I have when I try to read a biography of Bach and suddenly get lost in the musical notation. However much has happened in our understanding of both the science and life of Einstein in the last two decades, Pais did not choose to revise his book, which has left open the field. Indeed, I now turn to the three books that are the subject of this review.

Let me begin with the bad news. The bad news is a book called *Einstein's Daughter* by Michele Zackheim. This is easily the worst Einstein book I have ever read. The temptation is to ignore it, but since it has attracted some attention I feel obliged to include it. Let me make it clear from the outset that my objections to this book do not include the subject matter. Someday a competent, thoughtful person will write a good book about Einstein's *vie sentimentale*. We have learned in the last few years—since Pais wrote his biography—that it was much more complicated than any of us had imagined. Despite his disdain for the "merely personal," Einstein had a strong component of sensuality in his nature. It is not clear why one should find this surprising, any more than the fact that Isaac Newton died a virgin, but somehow we do. Each time there is a new revelation about Einstein it makes the headlines. And there will no doubt be more. There are sealed archives at the Hebrew University, some of which, it is reliably reported, have to do with this side of Einstein's life. They will be opened in 2006. Then it will be time for a serious book, hopefully written by someone with an understanding of Einstein's physics and an appreciation of the basic goodness of his character. Ms. Zackheim has neither.*

The ostensible subject of her book is her attempt to discover what happened to Einstein's daughter, whom he called with the diminutive Lieserl. We do not know her real name. The bottom line, which you discover after you have waded through some three hundred pages, is that Ms. Zackheim does not really know either. She has a guess—which I will come to—but the evidence for it borders on the ludicrous. In the meantime she has taken us on a wild goose chase through several countries—most notably Serbia—and has attempted to persuade us that Einstein was, if not a sexual predator, at the very least a male chauvinist pig. Let us begin with the agreed-upon facts. Einstein entered the Swiss Federal Polytechnical School—the Eidgenössische Technische Hochschule, the ETH—in October 1896. He was seventeen. Among his classmates was Mileva Marić (pronounced "Maritch"), who was three years older. She had been born in Titel, a town in southern Hungary that is now part of Serbia. Her father was a government official who somehow managed to accumulate a fair amount of money. It soon became clear that Marić had inherited a displaced left hip, relatively common in her family,

*On p. 79 of her book Zackheim writes, "Part of Einstein's *theory of relativity* involved disproving Newton's claim that light is deflected from [sic] the sun at 0.87 seconds of arc by proving that light is deflected by 1.7 seconds of arc." Leaving aside the fact that light is deflected *toward* the sun, the calculation cited by Zackheim was not done by Newton but by *Einstein*! He published it in 1911 in a paper that preceded general relativity, which came out in 1916.

which left one leg shorter than the other and apparently required her to wear an orthopedic shoe. I mention this only because, in the eyes of her family, it made her a difficult prospect for marriage, which probably helps to account for why they were more indulgent with Einstein than many parents might have been. It also helps to account for the fact that they were quite willing to let her go off on her own to continue her education in Switzerland. She had shown a considerable aptitude for physics and mathematics and was admitted to the ETH—one of the few places in Europe that allowed women to study these subjects—the same year that Einstein began his studies there. She was the only woman among a handful of men studying to become secondary-school teachers. It was inevitable that the two of them would meet.

They were antipodal personalities. Einstein had a great deal of self-confidence. He was attractive to women and had already developed a pleasant relationship with a girl he had met while in high school. Marić, probably because of her disability, was very shy. There was some history of mental illness in her family, and she had periods of serious depression. None of this seemed to matter to Einstein—at least then. He had fallen in love. His ardor apparently troubled Marić to the extent that she withdrew from the ETH and in the fall of 1897 went to Heidelberg where she audited courses. The first letter, so far discovered, between them was one she wrote to him from Heidelberg that fall. There are some fifty such letters that came to light as the result of an exhaustive search before the first volume of the collected correspondence was published in 1987. In addition, there were several hundred more that Marić wrote to friends and family. It is clear from the first letter that she was also thinking about a future with Einstein. She speaks of a present of tobacco from her father to him and expresses her hope that Einstein will visit her family. Until 1899 the letters are all written in the formal *Sie*; thereafter they became increasingly intimate with the use of *Du*. She had by this time returned to Zurich, and they were thinking of marriage, something that Einstein's parents—especially his mother—bitterly opposed. She objected to the age difference and the fact that Marić did not seem like a wifely type. She saw her as another student, like her son. Unspoken was the difference in religion—she was Serbian Orthodox and Einstein was a non-practicing Jew—and equally unspoken, one imagines, was the fact that his prospective bride was physically handicapped. Another factor was German (and in this case German-Jewish) contempt for the Slav. What was loudly spoken was the fact that neither of them had the money to support a marriage. Einstein graduated from the ETH in 1900 and was not able to get a real job—as a patent examiner—until 1902. This certainly had to do with both his outspokenness and anti-Semitism. He was barely eking out a living by giving private lessons in physics, along with some small help from his family. Marić, who had failed her exams at the ETH for the second time, had no means of support aside from her family. Given Einstein's general stubbornness and his self-image that he was something of a bohemian, I feel certain that if they had had the money they would have married—parental objections aside—before 1903, when they finally did.

In May 1901 Marić met Einstein at Lake Como in Italy for a few days of vacation, and she became pregnant. By July she had decided to return to her parents to present them with the news that she was going to have a baby. One of the very puzzling things about this is that Einstein did not accompany her. He had never met her parents. Perhaps he felt that this was not the moment. Or perhaps he did not have the money. He was certainly not going to tell his own parents about something that his mother had been warning him about from the beginning. At this remove it is very difficult to understand, and even more difficult to justify. He was certainly not indifferent to the birth of his daughter, which Marić's father announced to him the following January. I am going to give you two versions of Einstein's first known letter to Marić after the birth of the child—first the original, then Zackheim's cropped version. It is illustrative of her manipulation of quotations to buttress her case against Einstein. The letter is dated from Bern on February 14, 1902.

"My beloved sweetheart!

"Poor dear sweetheart, you must suffer enormously if you cannot even write to me yourself! [It seems to have been a difficult birth, and it was her father who had written.] And our dear Lieserl too must get to know the world from the aspect right from the beginning! I hope that you will be up and around again by the time my letter arrives. I was scared out of my wits when I got your father's letter, because I had already suspected some trouble. [This could be a reference to the fact that Mileva's displaced hip might make childbirth difficult. She did go on to have two other children, Einstein's sons Hans Albert and Eduard.] External fates are nothing compared to this. At once I felt like being a tutor with N[üesch] for two more years if this could make you healthy and happy. [J. Nüesch was a teacher of mathematics who had employed Einstein to work for him as a tutor. Einstein did not yet have his job at the patent office, and no university or laboratory was willing to give him one.] But you see, it has really turned out to be a Lieserl, as you wished. [He seemed to have hoped for a son.] Is she healthy and does she already cry properly? What kind of little eyes does she have? Whom of the two of us does she resemble more? Who is giving her milk? [her mother or a wetnurse?] Is she hungry? And so is she completely bald. I love her so much & I don't even know her yet! Couldn't she be photographed once you are totally healthy again? [So far, no credible photograph of Lieserl has ever been found.] Will she soon be able to turn her eyes toward something? Now you can make observations. I would like once to produce a Lieserl myself. It must be so interesting! She certainly can cry already, but to laugh she'll learn much later. Therein lies a profound truth. When you feel a little better, you must make a drawing of her. [If Marić did make such a drawing, it seems to have disappeared.]

"It is delightful here in Bern. An ancient, exquisitely cozy city, in which one can live exactly as in Zurich. Very old arcades stretch along both sides of the streets, so that one can go from one end of the city to the other in the worst rain without getting noticeably wet. The homes are uncommonly clean. I saw this everywhere yesterday when I was looking for a room. [Einstein had just moved to Bern, and all he could afford was a single room.] It does me extremely good to have escaped from the unpleasant environment. [He managed to rub most of his early employers the wrong way. He was too cocky.] I already saw to it that an advertisement will be published in the local gazette. I hope it will be of some help. If only I got two lessons a day I could save something for you. I have a large beautiful room with a very comfortable sofa. It only costs 23 francs. This is not much, after all. In addition, 6 upholstered chairs and 3 wardrobes. One could hold a meeting in it. Its plan follows. . . ."* Einstein then appends a carefully labeled drawing of everything in his room, down to the chamber pot. How a baby could fit into it is not specified. The room appears to be so small that one wonders how it could accommodate another person at all.

Now here is Zackheim's rendering of the same letter.

"Poor, poor sweetheart, you must suffer enormously if you cannot even write to me yourself! And our dear Lieserl too must get to know the world from this aspect right from the beginning! . . . I was scared out of my wits when I got your father's letter, because I had already suspected some trouble. I love her so much and I don't even know her yet!" To which Zackheim comments, "His letter [of which she has quoted only this fragment] suggests that not only did Mileva [have] a very difficult time, but that Lieserl suffered enormously too." Where in his letter does he "suggest" that "Lieserl suffered enormously"? This is typical

**The Collected Papers of Albert Einstein, Vol. 1.* The English Translation (Princeton, 1987), p. 191. For a different version, see *Albert Einstein/Mileva Marić: The Love Letters,* translated by Shawn Smith and edited by Jurgenn Renn and Robert Schulmann (Princeton, 1992), p. 73. Here is Smith's rendering of the first two sentences: "Poor dear sweetheart; what you have had to suffer if you can't even write me yourself anymore! It's such a shame that our dear Lieserl must be introduced to the world this way!"

Zackheim. There are much more flagrant examples that I will come to shortly. She takes some part of a document or some piece of questionable evidence and puts her own distorted spin on it without giving the unsuspecting reader a hint of what she is doing. This occurs on the next page when she describes Einstein's description of his room. "The room was much too large for him," she writes. " 'One could hold a meeting in it,' he joked. And yet he made no explicit mention of space for Lieserl." I invite the reader to examine Einstein's diagram of his room, with its chamber pot, which can be found in Volume 1 of his published correspondence, and decide if it is "much too large for him." Besides, there was an objective witness. Max Talmud, who had tutored Einstein when he was a boy in Germany, made a special visit to Switzerland at this time to see him. He reported that it was "a small poorly furnished room" that testified to "great poverty." It is said that poetry is what gets lost in translation. In Zackheim's case, it is balance. She has an agenda and allows nothing to get in its way.

For the next year Marić shuttled between Switzerland and her home. She was now on a resident visa, and the Swiss required that she leave the country every two months. Lieserl remained with her parents and, as far as we know, Einstein never saw her. On January 6, 1903, the couple married. Einstein had been working as a technical expert third class at the patent office with an annual salary of 3,500 Swiss francs since the previous June, so they now had enough money to live on. Their intentions toward Lieserl were never stated, but they seemed not to have made any attempt to bring her to Switzerland. This would have required Einstein to have escorted the child himself from Serbia so that she would appear on his passport and thus be legitimized. I do not know why he did not do this. Perhaps he thought there would be time later. By the following August, when Marić once again returned to her parents, the child was ill with scarlet fever. Marić reports this to Einstein, and he replies with concern. That is the last thing we know about her. After that, to all intents and purposes, she vanishes from the face of the earth. There is no birth certificate, no baptismal certificate, no adoption papers, no death certificate, and no grave site. Perhaps something will turn up someday, but what we realize after reading Zackheim's book is that she has not found anything solid either. What then does the book consist of?

It is largely a tale of wild and unproductive jaunts through war-torn Serbia accompanied by guides and translators. Zackheim does not speak Serbian, which does not prevent her from larding her text with proverbs such as "*Zaklela se zelmja raju da se tajne sve saznaju,* the earth pledged to paradise that all secrets will be revealed." This one occurs fairly early in the book and is spoken by a woman who assures Zackheim that they will find Lieserl. One of the things that Zackheim does, she tells us, is to pay off various potential informers with little presents of chocolate and money—even finding a job for one of them. Meanwhile we learn that there are rival scavengers on this treasure hunt. Zackheim barely beats one of them out, having uncovered no relevant information. As the book proceeds, it must have occurred to Zackheim, or her editor, that the ending was going to be an anti-climax. So the pace becomes more frenzied and the "revelations" more and more hysterical.

Near the end she begins quoting one János Plesch, a doctor who had also fled Germany. Plesch is notorious in the Einstein scholarly community as being a fabulist. Of him, Einstein once said, "Plesch is a swine, but he is my friend."* For a period in the late 1920s Plesch became Einstein's doctor, to the dismay of the

*I am grateful to Robert Schulmann for telling me this anecdote. For further information about Plesch, see "Some Reminiscences of Albert Einstein" by János Plesch and Peter Plesch, *Notes and Records of the Royal Society of London*, Vol. 49, no. 2 (1995). The article appears to have been largely written by Peter Plesch after the death of his father. A curious note in it is that the junior Plesch remarks that while he had confidence in his father as a doctor, he had less confidence in him as an historian. While not being a doctor myself, after reading this article I am not sure I have confidence in either. This article is the source for Zackheim.

medical community in Berlin who regarded him as a fashionable quack. In 1928 Einstein suffered a physical collapse and was diagnosed with a heart condition. Plesch ordered complete bed rest and a salt-free diet. This seemed to have worked. In the early 1930s both men left Germany, Plesch for England and Einstein for Princeton. Plesch had no further direct medical involvement with Einstein. But after Einstein's death Plesch, who concluded that Einstein was a strongly sexual person because of his full lips and large nose, proclaimed that he had had syphilis. This "diagnosis" was not based on any medical examination but from the fact that Einstein had died from an aneurysm and suffered from occasional bouts of anemia. Zackheim simply leaves this hanging, without any comment on its absurdity. Even more ridiculous is her claim—based, she says, on the alleged suspicions of his son Hans Albert, now conveniently dead, and the fact that the Einsteins had separate bedrooms and that Miss Dukas had a room near his study—that Dukas and Einstein had an affair. To bolster this contention she also claims that Dukas got the lion's share of Einstein's estate. I have tried to make it clear that Miss Dukas and Otto Nathan were *trustees* of the estate. Miss Dukas directly inherited only his books.* Anyone who spent five minutes with Miss Dukas would understand what a lunatic accusation this is. It made me wish for the resurrection of Dr. Nathan, for whom litigation was a blood sport. For him this book would have been a field of dreams. Miss Dukas died in 1982, so she is also conveniently absent. I was surprised that, in all her sniffing around under Einstein's bedsheets, Zackheim had not come across Marilyn Monroe's "list." This was a list that she and her roommate Shelley Winters discussed of celebrated men whom Marilyn wanted to sleep with. The list included Arthur Miller and Einstein. After Marilyn's death, Winters reports that a photo-

*On my visit to Einstein's house while Miss Dukas was still living in it, she showed me his study with his books. They were pretty serious—things like *The Golden Bough.* The collection did not look to me as if it had much material value.

graph of Einstein was found among her effects, signed "To Marilyn, with respect and love and thanks, Albert Einstein." I consider this to be as solid evidence for an affair between Einstein and Marilyn Monroe as anything that Zackheim uncovers.

What, then, are we left with? Here is the glorious climax of the book. In 1905 a Swiss polymath named Auguste Forel—he was also noted for his study of ants—published a book entitled (in English) *The Sexual Question: A Scientific, Psychological, Hygienic and Sociological Study.* Eventually it was translated into twenty languages. At some point Marić seems to have read this book. I say "seems" because the copy Zackheim is shown is alleged to have been hers. How she knows this I am not sure. I am also not sure how she knows that the underlined passages, of which she makes a great deal, were underlined by Marić. We also have no idea of when Marić (if she did) actually read this book. But for the sake of argument I will stipulate that the book and the underlinings were indeed Marić's. The question then becomes, do these underlinings, as Zackheim claims, somehow give us a clue as to what happened to Lieserl. Here are a few samples. I think they are fairly typical:

"If men are endowed with clear intelligence and an active mind, or with an intellectual or artistic creative imagination, they constitute excellent subjects for reproduction."

"Look for the indignation of parents when their children become betrothed to persons whom they consider to be beneath them in social position, or who possess too little money."

"The instinctive outburst of maternal love towards the newborn child corresponds to the natural right of the child, for the child needs the continual care of a young mother—so-called doting love. Nothing is so beautiful in the world as the radiant joy of a young mother nursing her child. And nothing, nothing is more degenerate than giving her child into strange hands."

All of these quotations could well apply to Einstein, Marić, and Lieserl, and none of them gives us a clue as to what really happened to her. Nonetheless Zackheim knows. She has it figured out. She uses the quotations like tarot cards. Her bout with scarlet fever left Lieserl, she tells us, "severely mentally handicapped." Einstein could not face up to this, so the child was abandoned somewhere in the folds of Marić's family. Zackheim then expresses her "belief" that Lieserl died on September 15, 1903, which coincided with a total eclipse of the sun! The mind reels. That there is no tangible evidence for any of this does not trouble Zackheim. She had a book to write.

Leaving this book for the other two on my list is like entering a parallel universe. In this other universe, conclusions are drawn from well-established premises. If there is an agenda, it is up front and the arguments for it presented explicitly. It is not the neon universe of Zackheim, it is the universe of real scholarship. The first of these books is *Einstein's German World* by Fritz Stern. Stern is a distinguished historian who is a professor emeritus at Columbia University. He has a deep understanding of Einstein's Germany because in 1938 he himself fled from it with his family. He was twelve. His father was a doctor. His uncle, Otto Stern, who won the Nobel Prize for Physics in 1943, had used his own money to come to Prague to work with Einstein in 1911, and then he followed him to Zurich before Einstein returned to Germany. These were people who were certain, until Hitler, that they had a lasting place in Germany. Einstein never had this illusion. He certainly did not foresee the Holocaust—who did?—but he had opposed German militarism from the time of World War I. When he left Germany in 1933 he was sure that he would never go back. Nothing changed his mind.

Thus Einstein's Germany really ended in 1933. If I have any complaint about Stern's book, it is the title. Many of the essays have very little to do with Einstein's Germany, and one of them, a critical review of Daniel J. Goldenhagen's *Hitler's Willing Executioners: Ordinary Germans and the Holocaust,* has nothing at all to do with Einstein. It seems to have wandered in from another book. But what makes this book essential reading for any student of Einstein is an essay of more than a hundred pages entitled "Together and Apart: Fritz Haber and Albert Einstein." If I could have waved a magic wand, I would have liked to see this essay expanded into a book. No one could have done it better. Haber was very close to Stern's parents and, indeed, was Stern's godfather. Who then was Fritz Haber? Haber was born in Breslau in 1868, making him a decade older than Einstein. He was one of Germany's greatest chemists. He won the Nobel Prize for Chemistry in 1918, for his technique for the synthesis of ammonia from nitrogen and hydrogen—something that is called the "fixation" of nitrogen—which led to the practical use of nitrogen for both explosives and fertilizers. In 1892 Haber, who had been born a Jew, converted and was baptized. This did not stem from a religious epiphany but from his conviction that his Jewishness was holding him back professionally. Haber had a relatively fragile ego and was constantly concerned about his professional status. He was also a German nationalist. At the beginning of World War I, Haber was one of the signers of the notorious Manifesto of the 93, which insisted that Germany's culture and its military tradition were inextricably mixed. Moreover Haber put his scientific genius to work creating poison gas. The introduction of the use of chlorine gas against Allied troops was largely Haber's doing. When Haber was awarded the Nobel Prize so soon after the war, this provoked an outcry that has some resemblance to the stir caused when Otto Hahn, one of the discoverers of fission, was awarded the Nobel Prize not many months after Hiroshima. One of the most horrible ironies of Haber's career, as Stern points out, was the development in his institute of Zyklon B as a possible pesticide. This became the gas used in the gas chambers. Since Haber died in 1934, he never knew this.

It must be clear from what I have already said that in nearly every way Haber and Einstein were anti-particles. Einstein had more trouble than Haber in finding suitable employment, one of the reasons for which, as Miss Dukas once wrote me, was "good old anti-Semitism." But even though Einstein was a nonpracticing Jew, it never occurred to him for a moment that he was anything but a Jew. When his "tribe"—as he called them—got into trouble in Germany, Einstein identified himself completely with the Jewish community. There is an incredibly moving photograph of Einstein in the early 1930s in a synagogue in Berlin. He is wearing a yarmulke and playing the violin in an effort to raise money for his fellow Jews. In World War I, Einstein was a pacifist. Needless to say, he did not sign the Manifesto of the 93. By the mid-1930s he had decided that force was the only way to stop Hitler. In World War II he did some consulting for the U.S. Navy: he worked on underwater explosives useful for destroying submarines. Einstein never felt German nationalism. In fact he never felt any nationalism. How then could he and Haber have possibly become friends and, indeed, very close friends?

In 1911 Haber became the director of the Kaiser Wilhelm Institute for Physical Chemistry and Electrochemistry in Berlin. In 1914 Einstein came to Berlin from Zurich. Along with a professorship, he had been appointed director of the Kaiser Wilhelm Institute for Physics, which was just being established. Haber, who had met Einstein a few years earlier, had been one of the prime movers in obtaining this position for him. Einstein was not entirely comfortable at first with the atmosphere he found in Berlin. It was much more formal than he was used to in Switzerland. Moreover his colleagues in physics, such as Max Planck and Max von Laue, did not seem to appreciate what he was trying to do. He was in the midst of creating his general theory of relativity and gravitation, which many people, including myself, think is the most magnificent edifice ever constructed in physics. But he was at that stage in the development where it looked pretty opaque. Even after he published it in 1916, most people found it pretty opaque. Only after the prediction of the theory that gave the precise amount of the bending of starlight by the gravitational field of the sun was confirmed in 1919, did the theory catapult into prominence. Einstein became one of the most celebrated people in the world. Whatever his faults, Haber apparently had a great gift for friendship. He tried to smooth Einstein's path into the social life of Berlin. But there was more to it than that.

Haber had two marriages, the first one being in 1901, when he married a converted Jewess named Clara Immerwahr. It was a troubled marriage from the beginning, and it remained troubled. In 1915, after what was apparently a violent argument which may have had to do with Haber's role with the military, Clara shot herself to death with his army pistol. In 1917 Haber remarried. His bride was twenty years younger than he was. That marriage ended in divorce in 1927. Thus Haber was someone who could lend a sympathetic ear to Einstein's own marital problems. Philipp Frank, who knew her, said that Marić was "blunt and stern," and added that "When he wanted to discuss with her his ideas, which came to him in great abundance, her response was so slight that he was often unable to decide whether or not she was interested." (Judging from his book, Frank did not like her much.) While Marić and their two sons did come with Einstein from Zurich to Berlin in 1914, the marriage was essentially over. The same year she moved back to Zurich with the boys.

Not to put too fine a point on it, Einstein essentially threw her out. What he did was to present her with a written set of demands—such things as keeping his clothes and laundry in repair—to which she would have to agree if she was to remain. He made it clear that this would be a business relationship and that all personal relations between them would cease. "You must desist immediately from addressing me if I request it," was one of the conditions. It is a dreadful document, but who knows what provoked it? At first Marić seemed willing to accept these conditions, but it became clear that they were the preliminaries to a divorce since Einstein had already begun a relationship with his cousin Elsa. So Marić left Berlin. In all of this Haber acted as an intermediary, someone whom both sides could confide in. It was a little like squaring the circle since Einstein was, above all, concerned not to lose contact with his sons, which he was sure would happen if they went back to Zurich with their mother. At the same time he was equally determined not to reconcile with her.

There were, of course, financial considerations. Much of Einstein's correspondence during these years concerns charges and countercharges as to how money was being spent. It is a curious commentary on human nature that, on the one hand, Einstein was creating modern cosmology and, on the other, insisting that Marić pay for tips. The actual divorce did not take place until 1919. In the divorce proceedings filed in Berlin, in which Einstein lists his religion as "dissenter," he admits to having had an adulterous affair with his cousin beginning in 1914. One of the stipulations of the divorce was that Marić would receive the proceeds of any eventual Nobel Prize. When Einstein won it in 1923, she received about $32,000. Over the years relations between them improved, and he was able to keep in touch with his sons—at least the older one, Hans Albert. The younger one, Eduard—"Tete"—had shown signs of mental instability from a very early age and died in 1965 in a psychiatric hospital in Switzerland. Hans Albert, who was educated at the ETH, became a professor of hydraulic engineering at Berkeley in 1947. He died in 1973. Marić never remarried. She died in Zurich in 1948.

When the eighth volume of the Einstein correspondence, the one that covers his Berlin years from 1914 to 1918, was published in 1998, a very strange letter emerged. It was written on May 22, 1918, by the then twenty-year-old sister of Margot Einstein, Ilse. The letter is written to a friend of the family named Georg Nicolai, to whom Einstein, as the collection shows, had written frequently. The letter was written with the instruction that it be destroyed immediately upon reading. Evidently it wasn't. In the letter Ilse tells Nicolai that Einstein has proposed marriage to *her*. She says that while she loves Einstein as a friend, she has no desire to marry him. She goes on to speculate on what ramifications such a marriage would have with her mother who, it will be recalled, has been having an affair with Einstein for some four years. It is quite unclear into what context to fit this letter. It certainly sounds sincere, but it had no effect on any subsequent events as far as I can see. Einstein married Elsa the following year, and in 1924 Ilse married a man named Rudolf Kayser, with whom Einstein seemed to have a most cordial relationship. It was Kayser who managed to save Einstein's papers in Berlin and arranged for their shipment by French diplomatic pouch to the United States. Ilse died in Paris in 1934.

As for Haber, by 1933 he understood what the coming of the Nazis meant—among other things, the end of his career in Germany. This was not as forgone a conclusion as one might think. Some German scientists of Jewish origin managed to remain in Germany by being hidden in industry, where the racial laws were not as rigorously enforced. One example was the Nobel Prize–winning physicist Gustav Hertz, also from Berlin. He was hired by Siemens and worked there throughout the war. Haber might well have found a similar refuge. But by this time he was thoroughly disillusioned with Germany. He resigned his post at the Kaiser Wilhelm Institute and emigrated. He died in Basel, Switzerland, in 1934. The next year Max Planck, despite objections from the regime, organized a memorial commemoration for him that was attended by five hundred people. It was almost the only instance when the German scientific community made a public demonstration against Hitler. One can only wonder what might have happened if more of them had spoken up earlier.

The last book I want to consider takes us from the profane to the sacred. It is *Einstein and Religion: Physics and Theology* by Max Jammer. Jammer is a professor emeritus of physics and the former rector of Bar-Ilan University in Israel. He is a philosopher of science and a historian as well as a physicist. He has written a number of important books about modern science, including *Concepts of Space,* with a foreword by Albert Einstein, with whom Jammer had several talks; *The Philosophy of Quantum Mechanics*; and, most recently, *Concepts of Mass in Contemporary Physics and Philosophy.* Jammer is a serious man who writes serious books. This one treats Einstein's outlook on religion and the reactions to it. There is also a discussion of attempts that have been made to use relativity and the quantum theory to buttress religious and mystical beliefs, something that I (and in this I seem to agree with Einstein) think is misguided. Einstein, as has already been noted, was a nonpracticing Jew. While this never changed, with the rise of Hitler his association with Judaism and Jewish causes increased. He became an early Zionist but eventually split with the Zionist leadership—such people as Chaim Weizmann—over the issue of the treatment of the indigenous Palestinian Arabs. Einstein felt that their interests were not being considered seriously enough.

However Einstein might have been viewed by others, he thought of himself as a profoundly religious person. But his religion did not include the concept of a personal God concerned with the working out of individual human destinies. His "God" was his faith that the universe was constructed according to some beautiful order and that the human mind would be able to comprehend it. He once wrote, "One may say the eternal mystery of the world is its comprehensibility." He often noted that his God was the God of Spinoza, the seventeenth-century Dutch-born philosopher. Spinoza also did not believe in a personal God. Jammer notes that what Spinoza did believe in sounds very much like what Einstein came to call "Cosmic Religion." For example, Spinoza wrote—in Jammer's translation from the Latin—"In the nature of things nothing is contingent, but all things are determined by the necessity of divine nature existing and operating in a certain model." Einstein believed that "God" put the universe together in a way that was unique and that the deepest science would reveal that architecture. Spinoza got himself into serious trouble for publishing his beliefs. He was excommunicated by the Jewish community. Einstein got into equally serious trouble when he published his.

By the late 1920s some religious spokesmen were claiming that Einstein's relativity theory was somehow "atheistic." In response to this, Rabbi Herbert S. Goldstein of New York cabled

Einstein in 1929, "Do you believe in God? Stop. Prepaid reply fifty words." Einstein cabled back, "I believe in Spinoza's God who reveals himself in the orderly harmony of what exists, not in a God who concerns himself with fates and actions of human beings." While this seemed to satisfy Rabbi Goldstein, it set off a firestorm. This intensified in 1930 when Einstein wrote an article for the *New York Times* adumbrating these views. It is difficult in this day and age to see why this provoked such a violent response. But it did. It lasted for years. Jammer gives many examples, of which one of the more temperate appeared in 1940, in the *Detroit Free Press.* Its editor, Malcolm Walker Bingay, wrote, ". . . to give up the doctrine of a personal God . . . shows that the good Doctor, when it comes to the practicalities of life, is full of jellybeans. . . ."

Einstein was concerned that people not misuse science as a foundation for religious belief. Science is empirical and transitory. It took just over two centuries for the Newtonian worldview to be replaced by Einstein's theory of relativity and gravitation. Newton certainly thought his theory expressed the inner workings of God. One does not know when or how relativity and the quantum theory will be replaced but, unless the law of scientific progress has been repealed, this will certainly happen. As to the question I started out with—has too much been written about Einstein?—my answer is "Not yet." But I also think that too much has not yet been written about love!

Heaven's Net: John Donne and Johannes Kepler

For of Meridians, and Parallels Man
hath weav'ed out a net, and this net throwne
Upon the Heavens, and now they are his owne.
—John Donne, "An Anatomie of the World—The First Anniversary"

What I am going to relate is the solution to a mystery. It is, I confess, a mystery that I might well have solved many years ago when I first came across it. I can only plead distraction. In any event, in 1983 I was sent for review by the *New Yorker* Daniel Boorstin's then newly published book *The Discoverers*. This was Boorstin's kaleidoscopic account of the history of scientific discovery from Aristotle to Einstein. I found the book somewhat maddening—a mixture of brilliant insights, incorrect statements, and paths not followed. Among the latter was a sentence that came just after Boorstin had quoted the well-known lines of

John Donne taken from his poem "An Anatomie of the World—The First Anniversary": "And new Philosophy calls all in doubt / The Element of fire is quite put out . . . ," which refers to the then new astronomical discoveries of Copernicus, Galileo, and Kepler. Boorstin casually remarks, "In 1619, when Donne visited the Continent, he took the trouble to visit Kepler in the remote Austrian town of Linz."

When I first read this laconic sentence I remember sitting bolt upright in my chair. Boorstin was announcing an encounter between arguably the greatest poet of the late sixteenth and early seventeenth centuries and unarguably one of the greatest astronomers who has ever lived. But that is all he does—just announce. No explanation, nothing. As I wrote in my review, "One longs to know what they talked about—even in what language. Kepler was always desperate to find converts to the Copernican system. What effect did the encounter have on the two men? Did it relieve the anxiety about the new cosmology which Donne suggests in the line [from the same poem] ' 'Tis all in peeces, all cohaerence gone . . .'?" In fact, did this unlikely encounter take place at all? How does Boorstin know? There are no specific references in his book. One is at a loss. This was the mystery, and I more or less forgot about it for the next twelve years.

But a few summers ago I happened to spot *The Varieties of Metaphysical Poetry* by T. S. Eliot in a local bookstore. This is the published version of the lectures that Eliot delivered in the 1920s and 1930s about the "metaphysical poets"—a term, incidentally, coined somewhat derisively by Dr. Johnson.* When I saw Eliot's

*The term was first used by Johnson, at least in print, in his *Lives of the Poets*. He gives the following wonderful characterization: "The metaphysical poets were men of learning, and to show their learning was their whole endeavour; but, unluckily, resolving to show it in rhyme, instead of writing poetry they only wrote verses, and very often such verses as stood the trial of the finger better than of the ear; for the modulation was the syllables." As absurd as this characterization seems to us, by the time of Johnson, Donne had fallen out of favor. Interest in his poetry did not really arise again until the beginning of this century.

book I immediately thought back to my old mystery—did Donne meet Kepler? But I soon found that on this Eliot was no help. Kepler does not appear in the index; neither does Copernicus, nor Galileo for that matter. The science in Donne's poetry was not Eliot's concern. But I was now determined to find out about this meeting. I decided that the next thing to do was to have a go at Kepler directly. The obvious way to start was to look at the Kepler entry in the *Dictionary of Scientific Biography.* Here I drew blood—a droplet. The Kepler entry, which runs to more than twenty pages, was written by Owen Gingerich of the Harvard-Smithsonian Center for Astrophysics. Gingerich, who is both an astronomer and an historian of science, has made many studies of Kepler. He is an expert and is to be taken seriously. Toward the end of his entry I found the sentence ". . . In 1619 the English poet John Donne had visited him [Kepler] in Linz, and in 1620 the English ambassador Sir Henry Wotton had called on him in Linz and had invited him to England." Since this was published in 1973 and since Boorstin's book was written in 1983, a reasonable assumption is that the latter had gotten his information from the former. But Gingerich, like Boorstin, offered no specific documentation. I had no more idea of the details of the visit than I had had before. Gingerich did, however, offer an extensive bibliography.

Two of the books in it seemed both relevant and accessible. The first was a biography written by the late Max Caspar, called simply *Kepler.* Caspar, who died in 1956 at the age of seventy-six, spent much of his life studying Kepler and was considered the leading scholar on the subject. Among other things, he was responsible for editing Kepler's complete oeuvre, including—important for what follows—his vast correspondence. The only other scientist of comparable ability that I know of who wrote as many letters as Kepler was Einstein. Caspar's biography is quite wonderful, and very readable, but of Donne there is not a word. On the other hand, Gingerich also cited Arthur Koestler's *The Sleepwalkers.* This is a fascinating book in which Koestler, with a novelist's eye, studies the characters of the scientists, including Kepler, who created the New Astronomy. There are several references to Donne, many of which make the interesting point that in Donne's prose polemic *Ignatius His Conclave,* Kepler and the other New Astronomers are mentioned by name, indicating that Donne was familiar with their work. But of the hypothetical meeting of the two men there is not a word. Given Koestler's interest in this kind of personal detail, one can only assume that when his book was published in 1963, no one knew about the meeting. That, it turns out, is correct. Something therefore must have happened between 1963 and 1973 when Gingerich made his reference.

This was as far as I was able to go with Kepler, at least using references that were available to me, so I decided to return to Donne. The standard biography of Donne was written by the late R. C. Bald.* Professor Bald did not live to finish his book, which was completed by W. Milgate of the Australian National University. It is a splendid biography, but there is only one mention of Kepler. This has to do with whether or not Donne read and was influenced by Kepler's science-fiction novella, *Somnium,* which was circulating in manuscript form during Donne's lifetime. It was published only after both Kepler and Donne had died. I do not find the evidence for such influence very convincing, but in any event there is in Bald's book no suggestion that the two men actually ever met. The book does make it quite clear, however, that the only time they *could* have met was in October 1619.

In February 1619 James Hay—Earl of Carlisle, Lord Doncaster—had been given a diplomatic mission on the Continent by King James I, to try to mediate the growing tensions between Catholics and Protestants in Middle Europe. The king had also appointed Donne chaplain to this mission. During the summer and fall of 1619 the mission wound its way eastward through Eu-

*In another fine biography, *Grace to a Witty Sinner,* written by Edward Le Comte (New York, 1965), Kepler is also mentioned in connection with the Somnium. There is no mention of any meeting.

rope in the quest of one last meeting with the recently anointed Holy Roman Emperor, Ferdinand of Styria, in the hope of enlisting his help in this mediation. Doncaster, who had met unsuccessfully with Ferdinand earlier, finally caught up with him again in early November in Graz, an encounter that Donne witnessed. From Bald's book we learn that Doncaster—and presumably Donne—before this encounter had sailed down the Danube from Germany to Vienna, where they arrived on November 2. They then proceeded to Graz. But Linz was on the way to Vienna. One would therefore guess from Bald's chronology that any meeting between Donne and Kepler had to have taken place before November 2, while the party was proceeding along the Danube to Vienna.

Now I was closing in. I had pinned down the approximate date of the meeting, and even the circumstances. But I was completely stuck. What to do next? It occurred to me to put my dilemma to a colleague. For a few years I had been exchanging information about the World War II German nuclear program with a young historian of science named Cathryn Carson. Carson did not work in the history of seventeenth-century astronomy, but she did get her degree at Harvard. Perhaps she had taken a course with Gingerich and the subject of Kepler and Donne had come up. I e-mailed her at Berkeley, where she was then teaching. She replied that she had never heard of the meeting, but she would e-mail Gingerich. Shortly thereafter she sent me a copy of an e-mail she had received from Gingerich. It had the missing reference! But it was something of a surprise. The reference was to a 1971 article entitled "Donne's Meeting with Kepler: A Previously Unknown Episode," which had appeared in the *Philological Quarterly,* written by a man named Wilbur Applebaum. I confess that neither the name of the author nor of the journal was familiar to me, though I later learned that it is a standard journal in its field and that Applebaum is a distinguished historian of science specializing in this period. Fortunately the Berkeley library had the journal, and in short order Carson faxed me a copy. The mystery was now about to be solved—I hoped.

Applebaum, who was then at the University of Illinois and is now a professor emeritus at the Illinois Institute of Technology, had made his discovery (I gathered from his brief three-page note) while perusing the edited version of Kepler's collected works, which Caspar had done in collaboration with Walther von Dyck and Franz Hammer. In Volume XVI Applebaum had discovered a letter that seems to have been overlooked by both Donne and Kepler scholars. He does not present the letter in full, or even say in what language it was written—German, it turns out—but he does give a few of the essentials. There is no date on the letter and the correspondent—who, though Applebaum does not tell us, turns out to be a woman—is unnamed. The editors, as Applebaum points out, have supplied the wrong date—namely 1608—as the date of the letter. The letter refers to an event which could only have taken place in 1619—Kepler's meeting with Donne and Lord Doncaster, spelled "Dancastre" in the letter. Donne is referred to as Doctore Theologo, which also helps to date the letter since he did not receive this honorary doctor's degree from Cambridge University until March 1615. The letter even fixes the meeting as having taken place on October 23, which fits the chronology adumbrated by Professor Bald.

I will not try to discuss the contents of the letter here but will come back to it when I explain, as nearly as I can tell, why there was such a meeting and what, if anything, were the consequences. Needless to say, once I learned of this letter I asked Carson if she could find it in the Berkeley library edition of Kepler's collected works, which she did. At the end of Applebaum's note he muses, "It would be interesting if Donne scholars could discover any reference in Donne's works or manuscripts to his meeting with Kepler." It would indeed. Certainly no one came forward in response to my query in the *New Yorker*. Now we shall reconstruct the "world-lines," the trajectories in space and time, which led to this extraordinary encounter on October 23, 1619.

Johannes Kepler was born on December 27, 1571, in the small Swabian town of Weil der Stadt in southwestern Germany. One of the pleasures in studying the life of Kepler is reading his own version of it. Until the present vogue of scientists, writing with increasing enthusiasm about what Einstein called with some scorn the "merely personal," no scientist—certainly no great scientist—has written about him- or herself with such candor. Of his father, a professional soldier who deserted the family, Kepler wrote that he was "vicious, inflexible, quarrelsome, and doomed to come to a bad end. . . . [He] treated my mother extremely ill, went finally into exile and died."* Kepler described his mother as "small, thin, swarthy, gossiping and quarrelsome, of a bad disposition." What Kepler could not have foreseen when he wrote this was that beginning in 1615 he would spend the next six years defending his mother from the charge that she was a *witch.* The denouement occurred in the fall of 1621, when under threat of torture his poor mother insisted that she had nothing to admit. She was freed but died a few months later.

Kepler was equally hard on himself. In the family horoscope Kepler produces a chronology which, as Koestler notes, reads like "the diary of Job." He notes that at age fourteen and fifteen he "suffered continually from skin ailments, often severe sores, often from the scabs of chronic putrid wounds in my feet which healed badly and kept breaking out again. On the middle finger of my right hand I had a worm, on the left a huge sore. . . ." This dreadful litany continues through 1592, when, at age twenty-one, he reports that "At Cupinga's I was offered union with a virgin; on New Year's Eve I achieved this with the greatest possible difficulty, experiencing the most acute pains of the bladder. . . ." It is interesting to note that at about the same time Donne, who was a few months younger than Kepler (his exact birth date is uncertain), was being described by a contemporary as a "great visitor of Ladies, a great frequenter of Playes, a great writer of conceited Verses." In fact it is likely that most of Donne's mag-

*These characterizations are part of a horoscope that Kepler drew up when he was twenty-six.

nificent love poems—"For Godsake hold your tongue, and let me love . . ."—were written before he was twenty-five—written, as he later said, by "Jack Donne" and not "Dr. Donne." Kepler was twenty-six when he produced his Job-like horoscope. If Donne had actually met Kepler at this time, he would certainly have thought him some sort of a Martian.

Kepler's early education was desultory. He was too sickly and frail to become an agricultural laborer. The only future his family could envision for him was the Lutheran ministry. As it happened, after the dukes of Württemberg converted to Lutheranism, they established a number of seminaries that could lead eventually to a state-sponsored education at the University of Tübingen, then to the theological faculty and the ministry. One of the features of the seminaries was their Latin. The curriculum was in Latin. Many students of Kepler's opus have remarked on the elegance of his Latin as compared to his German. There is something earthy about Kepler's German, even ungrammatical, as the letter that Applebaum turned up confirms.

From the beginning, Kepler's religious beliefs—which, along with his science, were at the core of his life—were always somewhat unconventional. He was a committed Lutheran, but he was more tolerant of other religions than was fashionable in an age of extreme religious intolerance. Throughout his life Kepler suffered terribly for his heterodoxy. One's heart goes out to him as time after time he is nearly destroyed because when it comes to religious principles Kepler will not budge an inch. In contrast, any student of Donne's life will be troubled by Donne's apostasy.* Donne was born into a very distinguished Catholic family; on his mother's side there was even a connection to Sir Thomas More. Donne was, at least as a young man, proud of his ancestry

*For a brilliant account of this and how it affected Donne's art, see John Carey, *John Donne: Life, Mind and Art* (New York, 1981). Professor Carey has some brief perceptive comments on Donne's reactions to the New Astronomy. He is apparently unaware of the fact that Donne and Kepler actually met.

and his religion. But being a Catholic at the end of the sixteenth century in England put one in mortal danger. One could be tortured to death in ways almost too unbearable to describe, especially if one was caught harboring a Jesuit or indeed any sort of Catholic priest. In 1593 Donne's brother Henry, who was a year younger, was arrested because he had had in his rooms a man suspected of being a priest. He would surely have been convicted of a felony except that he died from the plague while in jail awaiting trial.

For an English Catholic who did not choose to emigrate—as many did—there were three choices. One could practice one's religion clandestinely, hoping not to be caught. (During the reign of Queen Elizabeth, "recusants," as they were called, could practice their religion discreetly provided they also went to the Anglican service on Sundays and paid huge fines.) One could avow one's Catholicism openly and accept martyrdom, as some did. Or one could convert to the state Anglican creed. Donne did the first and then the last. When he matriculated to Oxford in 1584 his age was given as eleven. He was actually a year older, but at the age of sixteen, in order to obtain an Oxford degree, he would have been required to take an oath of allegiance to the Church of England. His modified age gave him an extra year at the university. As it was, he never graduated, and it is thought that he might have attended Cambridge for a few years after he left Oxford. But Donne was not a martyr. Indeed, his ambition was to become part of the English establishment. By 1597, when he became secretary to Sir Thomas Egerton, the so-called Lord Keeper of the Great Seal, the highest honor a lawyer could earn, who had himself converted and who was in a position to help Donne move up the governmental ladder, Donne had become an Anglican. This must have generated a lifelong agitation of his conscience. Donne's polemics against the Jesuits, who would not compromise, border on the irrational. There is something in what Dr. Johnson remarked a century later: "A convert from Popery to Protestantism gives up so much of what he has held as sacred as anything that he retains, there is so much laceration of mind in such a conversion, that it can hardly be sincere or lasting." "Lasting" Donne's conversion was. "Sincere" is another matter. When it came to religion, on the other hand, Kepler's conscience was clear and his outlook tolerant.

Kepler graduated from the Faculty of Arts at the University of Tübingen at the age of twenty. He was fortunate to have had as one of his teachers Michael Maestlin, an exceedingly competent astronomer. Maestlin taught Copernican astronomy, the doctrine that the earth moved around the sun and revolved on its axis every twenty-four hours, which Copernicus had published more or less on his deathbed in 1543, in his advanced classes which Kepler took. Kepler was grateful throughout his life for Maestlin's introduction to astronomy and above all to the work of Copernicus. He kept up a correspondence with his teacher, who was twenty years older but who outlived him, which Maestlin had more and more difficulty responding to. The situation is not unfamiliar. Maestlin was very good, but Kepler was a genius. In fact, what most of us learn in our general science classes as the "Copernican system" is, as we shall see, rather far removed from what Maestlin taught Kepler—that is, far removed from what Copernicus actually did. The modifications are due to Kepler. They represent radical departures from Copernicus's cosmology. They were too much for Maestlin. I think, incidentally, they were also too much for Donne. From the references to the New Astronomy in his poetry, it is my feeling, as I will explain, that Donne did not really understand it in detail.

After his graduation, Kepler matriculated at the theological faculty to begin his studies to become a Lutheran pastor. He had no thought of becoming an astronomer. His theological studies were to have ended in 1594, thence the ministry. But before that happened there occurred one of those twists of fate that seem to have been characteristic of Kepler's life and that kept propelling him toward his real destiny. The mathematics teacher at the Protestant seminary in Graz, Austria died. As happened frequently at the time, the seminary turned to the university in Tübingen in Germany to find a suitable replacement. Mathematical genius is pretty difficult to hide, and, as strange as Kepler may have seemed as a person, there was no doubt in anyone's mind that he was the most highly qualified individual to send to Graz. He spent the next six years there and indeed never returned to his native Germany except as a visitor.

On July 9, 1595, while teaching a class in Graz, Kepler had an epiphany. From our point of view it seems almost entirely mad. Indeed, it would hardly be worth mentioning in the present context except that just before Donne's visit in 1619 Kepler had revisited the subject and had written a book, *Harmonices mundi* (*Harmony of the World*),* which was, it appears, the main subject of conversation between the two men when they finally met. A significant difference between this book and the first, *Cosmographic Mystery,* which he published in 1596, was that the latter contained, along with the mystical detritus, an almost casual discussion of one of Kepler's greatest astronomical discoveries—the very nonintuitive connection between the length of time it takes a planet to complete one orbit around the sun (its period) and the average distance of the planet from the sun. This is called Kepler's Third Law. (We will deal with the first two later.) In mathematical language it states that the ratio of the square of the period of a planet to the cube of its distance from the sun is the same for all planets. It took Newton, a half-century later, to explain why it was true.

The question that Kepler posed in 1595 was in and of itself

*I am grateful to Professor Gingerich for reminding me that Harmonices is not to be translated as Harmonies—plural. He writes, "Kepler is being fancy, and because Harmonice is a Greek word, he uses Greek endings, so the 's' is in fact a singular Greek case." On the other hand, Professor Applebaum argues that *Harmonics of the World* is a better translation. He points out that the full title is *Five Books of the Harmonics of the World*—suggesting the plural. He also prefers "harmonics" to "harmonies." In the interest of harmony or harmonics, I present both points of view.

not unreasonable. He knew of six planets—Mercury, Venus, Earth, Mars, Jupiter, and Saturn. Pluto, Uranus, and Neptune had not yet been discovered. He wanted to account for the fact that there were *six* planets and for the five distances between them which had been roughly determined by Copernicus. Kepler knew—it had been known since Euclid—that there were *five* so-called regular solids, solids like the cube, which is bounded by six squares, and the pyramid, which is bounded by four equilateral triangles. He knew that a sphere could be put outside each of these solids so that all the vertices touched its surface. Similarly one can put a sphere inside each solid so that the sphere touches all its boundaries in the center. What Kepler tried to do was to put each planet on a sphere with one of the regular solids inside it and another outside it. This bizarre scheme did not work even on its own terms. He never could fit the planet Mercury. But Kepler was smitten with the idea. He sent copies of his book to every astronomer he could think of. Most important, for what happened later, he sent a copy to the Danish astronomer Tycho Brahe. Brahe was a great observational astronomer who had some not entirely realized theoretical ideas. He saw in Kepler, who was his junior by some twenty-five years, a mathematician of potential genius who might help him out.* The two connected sooner than either of them would probably have predicted. I believe—and I am not alone in this—that if Kepler had died just after the publication of his book (always a possibility in the disease-ridden environment in which everyone lived), he would simply have been an uninteresting historical curiosity. What made him such a great astronomer was the combination of his mystical urge to obtain ultimate laws with an absolute sense of empirical rigor. Kep-

*This substantially oversimplifies the situation. Brahe was engaged in a nasty priority fight with an astronomer named Ursus. When, eventually, he got Kepler's book, he thought he could use him to get at Ursus as well as using his computational talents. For a full account of these matters, see Victor E. Thoren, *The Lord of Uraniborg* (New York, 1990), the definitive biography of Tycho Brahe.

ler was never satisfied until whatever theory he had created fitted the known facts—at least as precisely as they were known.

In 1597 Kepler married for the first time. His wife, née Barbara Muehleck, was at the age of twenty-three already twice a widow. She, or her family, had pretensions, and it took some negotiation before Kepler, a poor schoolmaster, was allowed to marry her. One wonders why he did, since he described her as being "simple of mind and fat of body"—and worse. It does not seem to have been a very happy marriage, though it lasted fourteen years, until her death. She bore him five children, of whom two survived. In 1613 he remarried, to Susanna Rüttinger, an orphan of twenty-four, who bore him seven children, of whom, again, only two survived infancy or childhood.

Donne's marriage to Anne More, which took place in December 1601, was on the other hand the stuff of which operas are written. When he married her she was sixteen or seventeen and Donne twenty-nine. She had been living under the protection of Sir Thomas Egerton, the very Lord Keeper whose secretary Donne then was. She was the niece of Egerton's second wife, and the daughter of Sir George More, an influential political figure who was noted for his choleric temper. Knowing that her father would not approve of the marriage—Anne was still a minor—but being very much in love, the couple married secretly, thinking (naively as it turned out) that the old man might forgive them. When he was informed in February he went berserk. He had Donne arrested for marrying a minor and tried unsuccessfully to have the marriage annulled. He did succeed in getting Donne fired and, though he was released from jail, Donne was kept away from his wife until April.

After this episode Donne was never able to obtain another position with the government, though he tried repeatedly. He eked out a living for several years with support from patrons. King James, who seemed to take a special interest in Catholic converts if they had establishment connections such as Donne's, made it clear that he wished Donne to be ordained in the Anglican church, and that, indeed, would be Donne's only possibility for advancement. It is difficult to know Donne's real feelings about this, but in January 1615 he was ordained as deacon and priest in St. Paul's Cathedral and the same year given his honorary doctorate. It is also difficult to know his feelings about his marriage. His contemporary, Izaak Walton, remarked in his biographical sketch of Donne that Donne's marriage was the "remarkable errour [sic] of his life . . . ," though he added that "God . . . blest them with so mutual and cordial affections as in the midst of their sufferings made their bread and sorrow taste more pleasantly that the banquets of dull and low-spirited people." Before she died in August 1617, Donne's wife had given birth to twelve children! Five of them died either at birth or soon after. "Death be not proud . . ."

We can be sure that Donne read one book of Kepler's—*De Stella Nova* (Of New Stars), to use its common short title—which Kepler published in 1606. We can be sure because there is a marginal note to this effect in Donne's strange tract *Biathanatos,* which he wrote two years later. *Biathanatos* is an attempt by Donne to argue that committing suicide is not necessarily a sin. Why a tract on this subject should make reference to a book on "new stars" I will explain as soon as I have explained why Kepler's book was published in Prague. When we left Kepler he was living peacefully in Graz with his family. But increasingly the Counter-Reformation made the life of Protestants there more and more difficult. On July 27, 1600, the Archduke Ferdinand, a Catholic who ruled Graz, published a decree that demanded that everyone in the community submit to a public examination of their faith in four days. If they were not prepared to adopt Catholicism, they would be expelled from the city and their property confiscated. Accepting Catholicism for the sake of saving his job was something that Kepler was not prepared to do. Fortunately for him—and here is another example of his remarkable good fortune—in 1599 Tycho Brahe, after a dispute with the Danish authorities, had come to Prague to be the imperial mathematician, the official astronomer and astrologer for the emperor Rudolph. Kepler had visited Brahe just after his arrival and Brahe, desperate for mathematical help, now arranged a position for him. Thus Kepler came to Prague. Now to the new stars.

In 1572 Brahe had witnessed the explosion of a supernova in the constellation Taurus in our galaxy. Supernovae explosions in our galaxy are rare, hence it was remarkable—again, Kepler's luck—that there was a second explosion in September 1604 in the constellation Serpentarius. (The next one, as we have seen, was in 1987.) By this time Brahe had been dead for three years and Kepler had succeeded him. It was these new stars that were the subject of Kepler's book. But what was their relevance to Donne? Aristotle and his followers among the Schoolmen had maintained that the stars were made of a "fifth essence"—the "quinta essentia"—and not of earth, air, fire, and water. This essence was supposed to be indestructible and the stars made from it immutable. But the new stars showed that stars were in fact mutable. Aristotle and his followers in the schools were simply wrong. But if they were wrong about this, they could be wrong about anything, including a moral question like the ethics of suicide. This is the point Donne made. He writes in his *Biathanatos,* ". . . His [Aristotle's] Schollers [sic] stubbornly maintain his Proposition [the immutability of stars] still, though by many experiences of new Stars, the reason which moved *Aristotle* seems now to be utterly defeated." The New Astronomy was not merely a new science. It was a challenge to an entire moral and theological order. When Donne wrote, " 'Tis all in peeces, all cohaerence gone . . . ," he meant it.

There is one piece of writing by Donne that we are sure Kepler read. I will explain why shortly, after I describe what it was. In 1611 Donne published in both Latin and English the anonymous polemic satire which he called *Ignatius His Conclave.* Ignatius of Loyola, the founder of the Society of Jesus—the Jesuit order—had been canonized in 1622 by Pope Gregory XV. As I have mentioned, Jesuits reduced Donne to rage. In this satire

Donne imagines Ignatius in hell seated next to Lucifer, with whom he strikes up a companionship. This scene is witnessed by the "I" representing the author, who has had "liberty to wander through all places, and to survey and reckon all the rooms, and all the volumes of the heavens, and to comprehend the situation, the dimensions, the nature, the people, and the policy, both of the swimming Ilands, the *Planets,* and of all those which are fixed in the firmament." But having given a bold rendering of the furniture of the universe, the narrator backs off. He remarks, ". . . I thinke it an honester part as yet to be silent, than to do *Galileo* wrong by speaking of it, who of late hath summoned the other worlds, the Stars to come neerer to him, and give him an account of themselves." The preceding year Galileo, using the newly discovered telescope, had found new stars, the moons of Jupiter, and all sorts of new features on our moon. He had published a book, *Siderius nuncius* (Starry Messenger), describing these discoveries, which created a sensation. Donne had read the book, and a year later is making reference to it. But he goes on, "Or to Keepler [sic], who (as himselfe testifies of himselfe) *ever since* Tycho Braches [sic] *death hath received it into his care, that no new thing should be done in heaven without his knowledge.*" This shows that not only was Donne aware of Kepler, but he even knew that Kepler had inherited Tycho Brahe's position. But how do we know that Kepler read this? He tells us.

In 1609 Kepler wrote the first version of his science-fiction novella *Somnium, Sive Astronima Lunaris,* usually translated simply as *The Dream.* It concerns an imaginary voyage to the moon. The cast of characters includes the mother of the narrator Fiolxhilde. A few years later Kepler's own mother, who resembles the fictional Fiolxhilde, began her trials for witchcraft. Like many other science-fiction writers, Kepler was more concerned with describing the science of the moon than he was with the fiction. The original version of *The Dream* was not published but did circulate in manuscript. It has been claimed that Donne read it—as we shall see, Kepler thought so—but there is no clear evidence. In 1621, however, two years after Donne's visit, Kepler took up the *Somnium* again. This time he added 223 footnotes, which actually exceeded in length the original novella. The book was not published until 1634, four years after Kepler's death. Footnote 8 begins, "If I am not mistaken, the author of that insolent satire called *Ignatius His Conclave*, got hold of a copy of this little work of mine; for he stings me by name at the very beginning. As he goes along he brings poor Copernicus before the tribunal of Pluto. . . ." The first thing that strikes one about this quotation is why there is any doubt in Kepler's mind that the "author [Donne] of that insolent satire . . . got hold of a copy. . . ." After all, Donne had visited Kepler just a few years earlier. Why didn't Kepler simply ask him? The answer, it seems to me, is evident. Kepler did not know that it was Donne who wrote it, and Donne did not tell him. As far as I know, the first publication of *Ignatius* in which Donne was the acknowledged author occurred in 1652, some twenty years after the poet's death. When, in his letter (the one Applebaum turned up), Kepler described Donne's visit, no mention is made of *Ignatius*, nor indeed anything Donne wrote. In fact, one is led to wonder if Kepler had any idea of exactly who Donne was.

It has sometimes been claimed that because both *Ignatius* and *The Dream* involve voyages to the moon, Donne must have read Kepler before writing his polemic. I do not think so. In the first place, both men had such vivid imaginations that it is hard to understand why Donne would have *needed* to read Kepler to think of moon travel. He had read Galileo; that would have been inspiration enough. Besides, the two "voyages" are quite different. Kepler's travelers take an actual trip to the moon. They are carefully selected. Kepler writes, "No inactive persons are accepted into our company; no fat ones, no pleasure-loving ones, we choose only those who have spent their lives on horseback, or have shipped often to the Indies and are accustomed to subsisting on hardtack, garlic, dried fish, and such unpalatable fare. Especially suited are dried-up old crones who since childhood have ridden over great stretches of the earth at night in tattered cloaks on goats or pitchforks. The Germans are suitable, but we do not despise the lean hard bodies of the Spaniards." Kepler is careful to specify the distance to be traveled—fifty thousand "German miles," which is about the distance to the moon if one takes Kepler's definition of the German mile, which he gives in one of his footnotes. He is somewhat vague on the method of propulsion. He writes, "We congregate in force and seize a man of this sort; all together lifting him from beneath, we carry him aloft." And away they go!

Donne, on the other hand, describes a *potential* voyage to the moon. He employs a delightful conceit inspired by Galileo's telescope. The voyager can make "new *Glasses*" which will "draw the *Moone,* like a boate floating upon the water, as neere the earth as he will." One can just board the "boat." But, Donne goes on to say, it will be the Jesuits who will get on the now nearby moonship in order to establish a *"Lunatique Church"* on our celestial neighbor. On this subject Donne was relentless.

The "world-lines" of Donne and Kepler are now moving closer. But we still have to get Kepler to Linz from Prague and Donne to the European continent. Kepler remained in Prague from 1600 to 1612. His first few years there were the most productive of his scientific life. He transformed celestial mechanics forever. In doing so, in my view, he accomplished intellectual feats of overcoming that were comparable to what Einstein did when he created the theory of relativity. What did he have to overcome? A two-thousand-year *idée fixe* on how celestial bodies were allowed to move. From the ancient Greeks until Kepler, the only motion allowed to these bodies was uniform and circular. Some of the early astronomer-philosophers envisioned celestial objects as being attached to crystalline spheres, and some did not. But attached or not, the motions could only be uniform and circular. Unfortunately the actual motions of the planets, observed against the stars, are neither uniform nor circular. In fact the planets from time to time reverse their direction, something that is known as "retrograde motion." (The planet Mars, to take the example that Kepler focused on, moves eastward through the constellations for many months, then it slows down and comes to a stop. It then moves westward for several weeks before resuming its eastward trajectory.) Hence the problem that was posed to these savants was how to "save the appearances"—that is, how to reconcile the observed motions with the metaphysical demand that they be, in "reality," uniform and circular.

In the ancient world the most complete and successful attempt to "save the appearances" was made in the second century A.D. by the Alexandrine polymath Claudius Ptolemy. In the Ptolemic system the earth was stationary at the center of the universe. All the heavenly bodies, attached to spheres, revolved around the earth. But the motion of, say, a planet was not simply a uniform motion in a circle. There would be a sort of "guide point" that moved on a big circle. But around that the planet moved along a second smaller circle—a so-called epicycle. If that didn't work, one could add more circles. Or one could displace the earth from the center. A modern scientist confronted with this scheme understands immediately why, if pushed hard enough, it will work. What is happening is that the periodic motion of the planet is being analyzed in what is called a series of harmonics. We do this all the time. We can make the series as accurate as we want by including more and more terms—by adding more epicycles. In the Ptolemic scheme it required some thirty epicycles to reproduce the planetary motions to the degree of accuracy to which they were known. Copernicus changed the reference point from the earth to the sun—and, indeed, added *more* epicycles. In fact, if you look at an actual drawing of the Copernican system as developed by Copernicus, it will be a nest of circles on top of circles. It will bear almost no resemblance to the neat diagrams of the solar system you may remember from your general science course. Tycho Brahe complicated things still further by introducing a system in which the earth was at rest while the planets circled the moving sun. It was this chimera that Kepler had been engaged to work on. Fortunately, Brahe died in October 1601, before Kepler was forced to spend too much of his time on it—again, Kepler's luck.

Brahe and his assistants had assembled the most accurate data on the orbit of the planet Mars that had ever been taken. After Brahe's death, Kepler inherited the data and added some of his own. He also found a very ingenious way—using Mars as a reference point—to plot the orbit of the earth around the sun.* This had never been done before. It turned out that the earth's motion was not uniform either. But what was the orbit of Mars or, indeed, of any planet? After an incredible intellectual effort, by Easter of 1605 Kepler finally realized that it was an ellipse. You may remember from your high school geometry classes how to draw an ellipse. You take a string and tie down the ends with thumb tacks. Then you take a pencil and stretch the string and move it around a complete cycle to make your ellipse. The positions of the thumb tacks are called the "foci" of the ellipse. If you make the foci coincide, you have a circle, so that, in a sense, the ellipse is the simplest generalization of the circle. In Kepler's system—our system—the sun is at one of the foci of the orbital ellipse. Hence not only is the earth displaced from its central position, but the sun is not even at the center.

One is awed by the sheer force of Kepler's intellectual work.

*It should also be emphasized that unlike his predecessors—Brahe being a possible exception—Kepler dealt in *orbits,* that is, one-dimensional curves in space. Prior to this the planetary motion had been treated as the motion of the spheres. Brahe observed that comets showed no visible parallax so that they had to be beyond the moon and hence would have to intersect the spheres if they existed, which was one of the reasons he abandoned them. In Kepler's cosmology the spheres were also abandoned except for the stars. It is interesting that Galileo, Kepler's contemporary, did not accept the parallax argument and had notions about comets that appear to us to be quite absurd. He also thought that planets must have uniform circular motions. On the other hand, he understood the role of inertia in motions, which Kepler did not.

He had no computing machines. All the arithmetic was done by hand—hundreds of pages that still survive. He did not even have the mathematics he needed—the differential and integral calculus that would be developed later in the century by Newton and Leibniz. But, when he needed to, he invented approximation methods to a calculus that he did not even know existed. Kepler had a correct intuition as to how the whole mechanism worked. He thought it must be due to a force that emanated from the sun. The planets moved more rapidly when they were close to the sun since, Kepler reasoned, they were subjected to a greater force. Magnetism was then being studied, so he thought that magnetism might be the answer. It wasn't. As Newton pointed out, it's gravity. But it was Kepler who first understood that the motion around the planetary orbits was not uniform. Nonetheless he discovered a remarkable regularity. If you draw a line to the planet from the sun, in equal amounts of time this line (despite the nonuniformity of the motion) will sweep out equal areas. These two discoveries, along with the relation he found a decade later between the lengths of planetary periods and their distances from the sun, are known as Kepler's three laws. They were empirical discoveries but were guided by Kepler's intuition and his insistence that the planets should have the same physics as our own. It took the genius of Newton to show how they could be derived directly from the dynamics of gravitation.

Kepler put his elliptical discoveries into a book, *Astronomia nova* (to give its abbreviated Latin title, the *New Astronomy*), which he finally published in 1609. It can rightly be called the first truly modern astronomy book. I do not think Donne understood much of it, if he read it, which I also doubt. It is one thing to have a realization of the general implications of changing the locus of the solar system from the earth to the sun, or of discovering "new stars"; it is quite another to understand the sort of thing that Kepler did in the *New Astronomy*. Let me give a specific example from Donne's poetry, which shows me that he really did not understand in detail what had happened. My example is taken from the poem "A Valedictorian: Forbidding Mourning," which some Donne scholars think was written in 1611. The verse in question reads,

> Moving of th'earth brings harmes and feares,
> Men reckon what it did and meant,
> But trepidation of the spheares,
> Though greater farre, is innocent.

In the third line the use of "trepidation" resonates with "harmes and feares." But the rest of the line and the next show that Donne knew the astronomical usage of the term which had been current. The apparent motion of the sun, as measured in relation to the stars, is also not regular. Twice a year, at the so-called equinoxes, the orbit of the sun crosses the circle that lies between the two poles around which the stars rotate every 24 hours. This circle is called the celestial equator. But it was known ever since the ancient Greek astronomers that the equinox points themselves moved slowly—so slowly that it takes 26,000 years to complete one cycle. This they attributed to additional motions of the celestial spheres. Copernicus, on the other hand, attributed it to an additional motion of the *earth.* This motion is comparable to the periodic bobbing of a spinning top. Up to and including Copernicus, however, it was claimed that in addition to this so-called precession of the equinoxes there was an additional solar motion, a kind of oscillation which was called trepidation. Pre-Copernican astronomers accounted for this by a "trepidation of the spheares," a motion that one might think of as "innocent" since the spheres were made of the special celestial material. But Copernicus accounted for all the apparent motions of the sun against the stars by the earth's rotation about the sun and the revolution of the earth on its axis, including whatever weaving and bobbing might be needed to produce trepidation. There was in his system no "trepidation of the spheares" but rather a trepidation of the *earth.* Donne has missed this very essential difference. It is much less surprising that he missed Brahe's observational discovery that there is no such thing as trepidation. The earlier astronomers had simply been wrong. There is nothing to account for, although, despite Brahe's measurements which Kepler must have known about, Kepler still kept some slight irregularity in his analysis of the sun's motion. But this too would have been accounted for by a trepidation of the earth.*

Kepler's position in Prague was viable only as long as the emperor Rudolph remained in power. But the emperor showed increasing signs of mental instability and in 1611 was forced to abdicate. His brother Matthias succeeded him the following year. With the abdication of Rudolph, Kepler began looking for a more secure position. In 1612 he accepted one in Linz as the district mathematician and teacher in the district school. This was not a position with the distinction Kepler deserved, but he accepted it because he thought that his wife, who never liked Prague, might be happier there. But before she could make the move she died of typhus. Kepler remained in Linz, the capital of Upper Austria, until 1626—the longest he had ever lived anywhere. Now we have gotten Kepler to Linz. It remains to get Donne to the Continent.

A full accounting of the political and religious history that led up to the 1619 diplomatic mission on which Donne served—with its kings, queens, emperors, electors, dukes, archdukes, lords and ladies, ambassadors, bishops, popes, princesses and princes—would be, in my view, somewhat more complicated than, say, explaining the quantum theory. To keep things manageable I will focus on King James I of England and his two surviving children, Princess Elizabeth and her younger brother, Prince Charles. By the epoch we are considering, the rest of the king's offspring had died. King James, it may be recalled, was the son of Mary, Queen of Scots. This remarkable woman, who had been forced to abdi-

*I am grateful to Owen Gingerich for enlightening me on the fine points of Kepler's astronomy.

cate in favor of her infant son, was imprisoned in England and finally executed in 1578. When Queen Elizabeth I died in 1603, with no progeny of her own, James, who had been ruling Scotland as James VI, became James I of England. In the meanwhile he had married Princess Anne of Denmark. Indeed, one of the marriage ceremonies in 1589 was performed in Copenhagen. It was the king's only trip to the Continent, and, among other things, he visited Tycho Brahe's observatory. The mother of his bride, Sophia of Mecklenburg, then the queen mother of Denmark, was one of Brahe's patrons. As far as I can tell, this was the king's only venture into astronomy. Some years later he was presented with a signed copy of Kepler's book on the new stars. It is not clear that he ever read it.

This was an era when royal marriages were like moves in a diplomatic chess game. Hence when it came time for Princess Elizabeth and Prince Charles to marry, the choice of their mates depended a great deal on what ends the king wished them to serve. It was Princess Elizabeth's turn first. After some back and forth with other suitors, she was betrothed in 1613 to Frederick V, the Protestant elector palatine of the Rhine. In 1619, after the death of Emperor Rudolph's brother Matthias, Frederick served briefly as the king of Bohemia until he was deposed in 1620 by a Catholic army of Spaniards and Bavarians. He was sent into exile and died in obscurity in Germany. It was Frederick who was supposed to have protected the interests of Protestants in central Europe—something of concern to King James, who was also mindful of the well-being of his daughter.

King James was not exactly noted for his physical courage. At meetings of his court in England he wore a stiletto-proof vest. The king abhorred violence and, in particular, war. One of his notions for avoiding a war with Catholic Spain was to try to marry Charles off to a Spanish princess, the Infanta Maria, daughter of Philip III. James also wanted the Spanish monarch's help in restoring his son-in-law's position in Bohemia. The marriage never came to pass, much to the relief of most of the British people, who did not want their future king to marry a Catholic. He did anyway when in 1624 he signed a marriage treaty with Henrietta Maria, the daughter of Henry IV of France. King James died in 1625, and not long afterward Charles began warring on Spain. But in 1618, when the eruption of violence between Catholics and Protestants occurred in Bohemia, bringing with it a possible intervention from Spain, King James decided that he might be in a position to mediate between the feuding factions. To this end he contacted the king of Spain with an offer to make peace. The king of Spain, in turn, sought the advice of his ex-ambassador to England, the Conde de Gondomar, who replied with considerable candor, "The vanity of the present King of England is so great that he will always think it of great importance that peace should be made by his means, so that his authority may be increased. . . . [Nonetheless] it is possible and fitting to accept this mediation, since it cannot do any harm, or make things worse than they would be without it. . . . The King of England should send immediately to Germany an ambassador to treat about it. . . ." Thus the diplomatic mission on which Donne was to serve was born.

To lead what turned out to be an exercise in futility, the king chose James Hay, Lord Doncaster. Of Doncaster one historian writes, he was "a handsome Scot who by virtue of James I's susceptibility to young males, became one of the leading figures in the court and was created successfully Viscount Doncaster and Earl of Carlisle. Hay was best known for his facility in spending large sums of money, and under his guidance the vulgar excess which characterized Jacobean court entertainment reached a new height. A feast which he gave for the French ambassador at Essex House comprised sixteen hundred dishes which it took a hundred cooks eight days to prepare. . . ." Whatever enthusiasm Doncaster might have had for this mission was tempered by the fact that, as the queen had died in 1619, the entourage was required to wear mourning clothes. Donne had even less enthusiasm. His wife had died in 1617, leaving the surviving children in his care. He felt he was deserting them. For some reason Donne decided that he would never return alive from this trip. He wrote letters and preached sermons in the tone of a man condemned. He also announced that he would not be writing letters back to his friends in England. Hence we do not seem to have any record by Donne of his meeting with Kepler. Nonetheless the two men were, Donne above all, in no position to refuse a royal command.

Despite the mourning clothes, the mission was an extravagant one. It used more than twenty-five coaches for its ceremonial entry into Brussels. (Ultimately the mission cost about thirty thousand pounds, the equivalent of about three million dollars in present currency. To get some idea of what this sum meant, the *annual* salary of a domestic servant in a manor house was about three pounds.) There were lords and knights, a physician and cooks, and, of course, Donne, who was its chaplain. As I mentioned before, it meandered its way for some six months until it reached Linz on October 23, 1619. From Kepler's letter, which I am about to present, we can't tell who actually came to visit him—Donne alone, or Doncaster and some of his entourage. I would speculate that it was Donne alone. Nothing I have read about Doncaster leads me to believe that he would have had any interest in meeting Kepler. Furthermore, as far as I can tell, Kepler had no facility in Linz for entertaining such a group. In fact Kepler's letter suggests that he had seen Doncaster in town only from a distance. I would suppose that he and Donne spoke Latin. There is no reason to believe that Kepler knew English, or that Donne knew German. Until I read the letter I had hoped that the meeting might have been an epiphany—one of those occasions that changes the destiny of the participants forever. As you will see, it does not seem to have been. Here is the letter in translation. It is addressed to a woman. The German is somewhat rough-hewn* and in places not entirely clear. I have chosen what I think are the most plausible interpretations.

"Noble, much-honorable, virtuous, honorably-beloved sister in law: [This is very likely a form of address and not literally a reference to Kepler's sister-in-law. This sort of address must not have been uncommon. When late in life Donne wrote some of his most personal letters to his friend Anne Cokayne, he addressed her as "my noble dear sister."] I received your letter and report from England a month ago. From it I gathered your and your honorable cousin Burlemachi's favorable efforts and diligence in presenting my work."

The work in question is certainly Kepler's *Harmony of the World,* which he had published in the summer of 1619. His intent was to dedicate the book to King James, in part as recognition of his role as a peacemaker. Kepler was no doubt also aware that the king's son-in-law, Frederick, was the protector of the Protestants in Bohemia. But the sheet with the dedication was not delivered until early in 1620, after Donne's visit. By this time Frederick was in serious trouble, and many of the printed copies do not have the dedication.

The letter goes on, "And because I now also reply confidently on the continuation of this promotion, I cannot help reporting again (perhaps my last letter did not get through) that I have talked with a Doctor of Theology named Donne, who was traveling with His Royal Majesty's envoy, Mr. Doncastre [sic], and appeared here on October 23. [This would seem to imply that there was another letter dealing with the meeting that may not have been preserved.] I told him that I had ordered the dedication copies [presumably the copies of *Harmony of the World* dedicated to King James], and added that, because I had seen the envoy

*I am very grateful to Gerald Holton and Freeman Dyson for help in translating this letter. Both of these scholars remarked on the earthiness, even the apparently ungrammatical nature, of Kepler's German. As I have noted, this was in sharp contrast to his Latin, which was elegant.

[Doncaster] here [in Linz], wanted to ask his Grace to convey and commend the work. Because one could sense that this wish was well received, and because Mr. D. Donne [sic] thought that the envoy would return to court before the presentation [of the dedication] copies, and since this promised every beneficial promotion, I wrote on his advice, a letter in French to Mr. Burleranchi [sic] [Professor Applebaum has informed me that "Burleranchi" or "Burlemachi" was a gentleman whose actual name appears to have been Burlemaqui. He was of Italian-Protestant descent but had been naturalized English. He seems to have been a somewhat shady "wheeler-dealer"—he was once fined ten thousand pounds by the Star Chamber—but was used by the crown in some of its international financial dealings. Kepler apparently believed that Burlemaqui might be an entrée to the king.] and gave it to the doctor. In this letter I informed him [Burlemaqui] about all this and, in case the presentation [to the king] had not happened yet, asked him to take note of Mr. D. Donne's and the envoy's opinion about it.

"But because I have now learned that the envoy will arrive long after the presentation of the work [in fact the mission did not return to London until January] I am concerned that my earlier letter might keep Mr. Burlemachi from promoting this work. And because my gracious sister-in-law might have some thoughts about this, I would ask you in the meantime to write letters supporting it, still showing respect to the envoy so that what you do would not be interpreted as fickleness on my part." I have not translated the last paragraph of the letter, which has to do with the financial aspects of printing the book.

It seems to me clear from this letter, and from what has gone before, that as far as Kepler was concerned, Donne was merely an appendage to the British mission who might be useful in promoting Kepler's work in England. Of the "Jack Donne" who wrote some of the greatest poetry in the English language, he knew nothing—still less of the Donne who wrote *Ignatius* or who preached upon the tolling of the bell. A year later Sir Henry Wotton, who had been with Donne as a student at Oxford and remained one of his closest friends throughout his life, visited Kepler in Linz. Wotton, who had had a long career as a diplomat in Venice, was temporarily functioning as King James's envoy in Middle Europe. Some years earlier Wotton had been in serious difficulties with the king because, as a joke, he had written in an album of remembrances belonging to a German friend, "An Ambassador is an honest man, sent to lie abroad for the good of his country." Somehow the king learned of this and was not pleased. In time Wotton was forgiven and in 1620 came to Linz to offer Kepler a job in England. My guess is that during his conversation with Donne, Kepler must have conveyed some of his unhappiness in Linz. From the time he arrived there he found himself in conflict with the orthodox Lutherans in the community and was even denied communion. In 1616 there was an attempt to eliminate his job altogether. Nonetheless, when it came to actually relocating abroad, Kepler could not make himself do it.* He stayed in Linz until 1626, when the flames of war literally consumed the city. From that time until his death in 1630, he found no tranquil place where he could do his work. He died in Regensburg, Germany, on November 15. He was buried in the churchyard in front of Peter's Gate. The grave and the churchyard were swept away in the course of the Thirty Years' War. Donne died some six months later, on March 31, 1631. He was buried in St. Paul's Cathedral. A statue in effigy was carved to mark his resting place. Of it his friend Sir Henry Wotton remarked, "It seems to breath faintly; and, posterity shall look upon it as a kind of artificial Miracle."

*One wonders how he might have fared in England, indeed what job he could have been offered. Court astrologer was not on the list since King James was wont to put astrologers who did *his* horoscope in prison. This actually happened to Thomas Harriot, a scientific polymath, who had had a brief correspondence with Kepler on optics.

The Merely Very Good

"I learned from my father that it was silly to want to be Lloyd George, but how can I learn that it is silly to want to be Beethoven or Shakespeare? I shall have to be all I can be in order to learn this."—Stephen Spender, Journal entry, October 26, 1939

"Being a minor poet is like being a minor royalty, and no one, as a former lady-in-waiting to Princess Margaret once explained to me, is happy as that."—Stephen Spender, Journal entry, October 5, 1979

"Nevill Coghill, W. H. Auden's tutor at Oxford in 1926, asked what he—Auden—wished to be later in life. 'I'm going to be a poet,' Auden answered. Coghill made some patronizing reply ('Ah yes! . . . It will give you insight into the technical side of your subject') only to be crushed by Auden: 'You don't understand at all. I mean to be a great poet.' "—Richard Davenport-Hines, *Auden*

Early in 1981 I received an invitation to give a lecture at a writers' conference which was being held someplace in Pennsylvania. I don't remember the exact location, but a study of the map persuades me that it was probably New Hope on the Delaware River, across from New Jersey. It was certainly on the Delaware River. My first inclination was to say no. There were several reasons. I was living in New York City and teaching full time. My weekends were precious, and the idea of getting up before dawn on a Saturday, renting a car, and driving across the entire state of New Jersey to deliver a lecture was repellent. As I recall, the honorarium offered would have barely covered my expense. Furthermore, a subject had been suggested for my lecture that in truth no longer interested me. Since I both wrote and did physics, I had often been asked to discuss the connection—if any—between these two activities. When this first came up I felt obligated to say something, but after twenty years, about the only thing I felt like saying was that both physics and writing—especially if one wanted to do them well—were extremely difficult.

The conference seemed to be centered on poetry, and one of the things that came to mind was an anecdote that J. Robert Oppenheimer used to tell on himself. Since Oppenheimer will play an important role in what follows, I will explain. After Oppenheimer graduated from Harvard in 1925—in three years, summa cum laude—he was awarded a fellowship to study in Europe. Following a very unhappy time in England, where he seems to have had a sort of nervous breakdown (he physically attacked an old friend, Francis Fergusson, following an intimate discussion they had about Oppenheimer's feelings of sexual inadequacy), he went to Germany to get his Ph.D. He studied with the distinguished German theoretical physicist Max Born in Göttingen and took his degree in 1927 at the age of twenty-three. Born's recollections of Oppenheimer, which were published posthumously in 1975, were not sympathetic. He wrote, "[Oppenheimer] was a man of great talent, and he was conscious of his superiority in a way which was embarrassing and led to trouble. In my ordinary seminar on quantum mechanics, he used to interrupt the speaker, whoever it was, not excluding myself, and to step to the blackboard, taking the chalk and declaring: 'This can be done much better in the following manner. . . .' " In fact, it got so bad that Oppenheimer's fellow students in the seminar petitioned Born to put a stop to it.

Quantum mechanics had been invented the year before by Erwin Schrödinger, Werner Heisenberg—also a Born disciple—and Paul A. M. Dirac. The next year Dirac came as a visitor to Göttingen and, as it happened, roomed in the large house of a physician named Cario where Oppenheimer also had a room. Dirac was twenty-five. The two young men became friends, in so far as one could have a friendship with Dirac. As young as he was, Dirac was already a great physicist, and I am sure he knew it. But he was, and remained, an enigma. He rarely spoke, and when he did it was always with extraordinary precision and often with devastating effect. This must have had a profound impact on Oppenheimer. While Oppenheimer was interrupting Born's seminars, announcing that he could do calculations better in the quantum theory, Dirac, only two years older, had *invented* the subject! In any event, in the course of things the two of them often went for walks. In the version of the story that I heard Oppenheimer tell, they were walking one evening on the walls that surrounded Göttingen and somehow were discussing Oppenheimer's poetry. He had published a poem in *Hound and Horn.* I would imagine that the "discussion" was more like an Oppenheimer monologue, which was abruptly interrupted by Dirac who asked, "How can you do both poetry and physics? In physics we try to give people an understanding of something that nobody knew before, whereas in poetry . . ." Oppenheimer allowed one to fill in the rest of the sentence.

Weighted against these excellent reasons for my not going to the Delaware River conference were two others that finally carried the day. In the first place, I was in the beginning stages of a love affair with a young woman who wanted very much to write. She wanted to write so much that she had resigned a lucrative job with an advertising agency, taken her savings, and was giving herself a year in which she was going to do nothing but write. It was a gutsy thing to do, but like many people—most, no doubt—who try it she was finding it pretty rough going. In fact, she was rather discouraged. So, to cheer her up, I suggested attending this conference where she might have a chance to talk with other people who were in the same boat. This aside, I had read in the tentative program of the conference that one of the other tutors was to be Stephen Spender. This, for reasons I will now explain, was decisive.

I should begin by saying right off that I am not a great admirer of Spender's poetry. He is, for me, one of those people whose writing about their writing is more interesting than their writing itself. I had read with great interest Spender's autobiography, *World Within World,* especially for what it revealed about the poet who *did* mean the most to me, namely, W. H. Auden. Auden's Dirac-like lucidity, the sheer wonder of the language, and the sense of fun about serious things—"At least my modern pieces shall be cheery / Like English bishops on the Quantum Theory"—were to me irresistible. I had also spent an afternoon with Auden under circumstances I will shortly describe. That too involved Oppenheimer. To me, Spender was a last link with that great generation of British poets that by 1981 had almost all died off. What I did not know in 1981—I learned it only after Spender's journals were published in 1986—was that Spender had paid a brief visit to the Institute for Advanced Study in Princeton in November 1956, the year before I got there, and the year before I had been the instrument in getting Auden there for an equally brief visit.

Spender's journal entry on his visit is fascinating, both for what it says and what it does not say. He begins by noting that "Oppenheimer lives in a beautiful house, the interior of which is painted almost entirely white." This was the director's mansion. Spender did not notice that because of Oppenheimer's Western connections, there was also the odd horse on the grounds. He goes on, "He has beautiful paintings. As soon as we came in, he said: 'Now is the time to look at the van Gogh.' We went into his sitting room and saw a very fine van Gogh of a sun above a field almost entirely enclosed in shadows." At the end of my first interview with Oppenheimer, immediately after I had driven cross-country from Los Alamos in a convertible with a large hole in the roof and had been summoned to the interview while still covered in dust, he said to me that he had some pictures I might like to look at some time. Some months later I was invited to a party at the Oppenheimers and realized that he was talking about a van Gogh. Some years later I learned that this was part of a small collection he had inherited from his father, to which he had never added. A few days after Oppenheimer's funeral in 1967 I was at the Pilates gymnasium in New York where George Balanchine was also a client. I had noticed that Balanchine was listed on the funeral program as having selected the music for it. I said to Balanchine that I didn't realize he was a friend of Oppenheimer's. Balanchine said he wasn't, but a few days before Oppenheimer's death he had been invited to Princeton where he and Oppenheimer, who knew that he was dying, discussed the music to be played at the funeral—*Requiem Canticles* by Stravinsky and Op. 131, the Quartet in C Sharp Minor by Beethoven. Balanchine was at a loss to explain this incident. I recall telling him about the van Gogh.

Spender then goes on to describe Oppenheimer's appearance: "Robert Oppenheimer is one of the most extraordinary-looking men I have ever seen. He has a head like that of a very small intelligent boy, with a long back to it, reminding one of those skulls which were specially elongated by the Egyptians. His skull gives an almost eggshell impression of fragility, and is supported by a very thin neck. His expression is radiant and at the same time ascetic." Much of this description seems right to me except that it leaves out the fact that Oppenheimer did have the sun-wrinkled look of someone who had spent a great deal of time out of doors, which he had. He also does not seem to have remarked on Oppenheimer's eyes, which had a kind of wary luminescence. Siamese cats make a similar impression. But more important, Oppenheimer appears in Spender's journal completely out of any contextual relevance to Spender's own life. There is no comment on the fact that, three years earlier, Oppenheimer had been "tried" for disloyalty to the United States and that his security clearance had been taken away. One of the charges was that his wife, Katherine Puening Oppenheimer, was the former wife of Joseph Dallet, who had been a member of the Communist party and who had been killed in 1937 fighting for the Spanish Republican Army. In 1937 Spender was also a member of the Communist party in Britain and had also spent time in Spain. Did Oppenheimer know this? He usually knew most things about the people that interested him. Did "Kitty" Oppenheimer know it? Did this have anything to do with the fact that during Spender's visit she was upstairs "ill"? Spender offers no comment.

A year later it was Auden's turn. For this I was responsible—at least catalytically—and I will now explain why. It goes back to my undergraduate days at Harvard. The Harvard chapel had Sunday services which were open to the public and which had guest ministers. On my father's advice—he knew him—I went to hear Reinhold Niebuhr. Niebuhr was the most electrifying orator I have ever listened to. He spoke in the straightforward language of the blue-collar parish he had ministered to in Detroit, but with an eloquence and profundity that was riveting. When he remarked, with a piercing look that I was sure was beamed directly at me, on the dangers of a "neurotic preoccupation with self," I felt that somehow he had been able to look directly into my soul. I wanted very much to meet him. Much to my amazement, when I saw the list of the Institute for Advanced Study for 1957–1958, there was Niebuhr. I told this to my father, and he said by all means go and see Niebuhr and give him his greetings.

As it happened, the Niebuhrs lived close by in a new Institute housing complex. Summoning up my courage, I went over after dinner one evening and knocked on their door. It was answered by Niebuhr's wife Ursula, an Englishwoman of astounding charm and vitality. I explained who I was, and she invited me in. I then had a very nice chat with Niebuhr. I learned later from his wife that he had been feeling rather depressed, partly due to some medication he was taking for a heart condition and partly due to the fact that he felt somewhat left out of the intellectual ferment at the Institute. At the time this was centered on some astonishing new discoveries in physics, especially those involving the breakdown of parity symmetry. T. D. Lee and C. N. Yang were at the Institute that fall. That year they won the Nobel Prize. I was able to fill Niebuhr in and to sort out for him the various players in the game. I was encouraged to visit on a regular basis, and I did.

In the course of one of these visits the subject of Auden came up. Perhaps I had just read one of his poems. It was as if I had caused a geyser to explode. I had no idea that Ursula Niebuhr was one of the women whom Auden was close to. They had known each other since some earlier Oxford time. When she pronounced his first name, somehow "Wystan" became "W*h*iistan"—like "whisk." I also had not known of Auden's admiration for Niebuhr's writing. He had dedicated his collection of poems, *Nones,* to the Niebuhrs. It turned out that Auden had a visiting appointment at Princeton University. He did not realize that the Niebuhrs were in Princeton, and they did not realize that he was at the university. I was able to provide the link when accidentally I found myself on a train from Princeton to New York, seated next to Auden. Upon returning to Princeton I told the Niebuhrs about this, and Ursula was ecstatic. She said that she would at once invite Auden for tea, and then she would invite me.

I had more or less forgotten about the matter when, a few days later, I received an invitation from the director's office—conveyed on behalf of Oppenheimer and Ursula Niebuhr—to meet there in advance of a luncheon that was to be held in Auden's honor in the Institute cafeteria. No van Gogh this time. Upon arriving at the office I found the Niebuhrs, Auden, the historian Sir Llewellyn Woodward and his wife, and Oppenheimer and Kitty. After a certain amount of hemming and hawing, we all trooped to the cafeteria.

My recollections of what followed are somewhat spasmodic. I do recall the group of us marching into the bustling cafeteria and my getting some pretty strange glances from my colleagues. They must have wondered what on earth was going on. I also recall the seating arrangement at the round table that Oppenheimer had claimed for us. Niebuhr was to my right and Ursula to my left. Auden was seated next to her. Oppenheimer was across the table. The Woodwards, both of whom had that slightly mottled look one sometimes finds in elderly British intellectuals, said absolutely nothing during the lunch. Kitty made a few Delphic remarks. Oppenheimer told a couple of anecdotes—the Dirac story, as I recall, and a rather pointless one about his attempt to learn Sanskrit in Berkeley.* Auden's unkempt face had a look of inscrutable bemusement. He spent the rest of the lunch gossiping with Ursula Niebuhr. All in all, he couldn't have been more pleasant. From time to time, however, Niebuhr caught my eye with a look that suggested that "this too shall pass." It did, and after lunch I took Auden on a tour of the Institute. We stopped in to see Freeman Dyson, and he and Auden played some

*The following year Dirac was a visitor at the Institute. I witnessed the following. He was scheduled to give a lecture in New York. A reporter had heard about it and phoned him for an advance copy. Since Oppenheimer had decided that we couldn't have phones in our offices for fear that we would be distracted, there was instead a hall phone that distracted everyone, so that I could hear Dirac's negotiations with the reporter. As it happened, Abraham Pais was in my office, and after Dirac was finished he came in to ask Pais's advice. What should he write on the copy of the lecture so that it wouldn't be misquoted. Pais suggested, "Do not publish in any form." For several minutes Dirac stood there in absolute silence while Pais and I continued our conversation. Then he said, "Isn't 'in any form' redundant in that sentence?"

sort of complex crypto-analytic word game on Dyson's blackboard. I think they were having a good time. I walked Auden in the general direction of the Princeton train station and never saw him again.

In the fall of my second year at the Institute, Dirac came for a visit. We all knew that he was coming, but no one had actually encountered him despite rumored sightings. By this time Dirac, who was in his mid-fifties, had a somewhat curious role in physics. Unlike Einstein, he had kept up with many of the developments and indeed, from time to time, commented on them. But, like Einstein, he had no school or following and had produced very few students. He had essentially no collaborators and once, when asked about this, had remarked that "the really good ideas in physics are had by only one person." That seems to apply to poetry as well. He taught his classes in the quantum theory at Cambridge University, where he held Newton's Lucasian chair, by literally reading in his precise, clipped way from his great text on the subject. When this was remarked on, he replied that he had given the quantum theory a good deal of thought and that there was no better way to present it.

At the Institute we had a weekly physics seminar over which Oppenheimer presided, often interrupting the speaker. Early in the fall we were in the midst of one of these—there were about forty people in attendance in a rather small room—when the door opened. In walked Dirac. I had never seen him before, but I had often seen pictures of him. The real thing was much better. He wore most of a blue suit—trousers, shirt, tie, and, as I recall, a sweater—but what made an indelible impression were the thigh-length muddy rubber boots. It turned out that he was spending a good deal of time in the woods near the Institute with an ax, chopping a path in the general direction of Trenton. Some years later, when I began writing for the *New Yorker* and attempted a profile of Dirac, he suggested that we might conduct some of the sessions while clearing this path. He was apparently still working on it.

The profile was never written since eventually Dirac decided that he didn't want to be bothered, but one thing remains with me. He told me that his habit of silence began in his childhood, when his father insisted that the family speak French at the table. Since Dirac did not know much French, he remained silent and continued to do so. But here he was, muddy boots and all, in the middle of our seminar, his greying hair somewhat touseled by his efforts in the woods. Oppenheimer found him a seat. He sat down and looked around at all of us with a faint smile. A colleague sitting next to me whispered in my ear, in an imitation of Dirac's voice, a variant of that line from the old British joke, "Is this what the common people call physics?" I think we all felt a bit silly. I don't know how Oppenheimer felt.

Now it is some twenty-five years later. The sun has not yet come up and I am starting a drive across the state of New Jersey with my companion. We have left New York at about 5 a.m. so that we can have breakfast en route and still arrive in time for my 9:30 talk. I have cobbled something together about physics and writing. As we go through the tunnel I recall an anecdote T. D. Lee once told me about Dirac. He was driving him from New York to Princeton through this same tunnel. Sometime after they had passed it, Dirac interrupted his silence to remark that, on the average, about as much money would be collected in tolls if they doubled the toll and had tollbooths only at one end. A few years later the New York Port Authority seems to have made the same analysis and halved the number of tollbooths.

We pass the turnoff that would have taken us to Princeton. It is tempting to pay a visit. But Oppenheimer is dead, and Dirac is living in Florida with his wife, the sister of fellow physicist Eugene Wigner. Dirac used to introduce her to people as Wigner's sister, as in "I would like you to meet Wigner's sister." Dirac died in Florida in 1984.

We arrived at the conference center a few minutes before my scheduled time. There was no one—or almost no one—in the lecture room. In mid-room, however, there was Spender. I had never met him, but I recognized him at once from his pictures. Isherwood once described Spender's eyes as having the "violent color of blue-bells." Spender was wearing a dark blue suit and one of those striped British shirts—Turnbull and Asser?—the mere wearing of which makes one feel instantly better. He had on a club tie of some sort. He said nothing during my lecture and left, with the minuscule audience that I had traveled several hours by car to address, as soon as it was over. My talk was followed by a mediocre lunch in one of the local coffee shops. There seemed to be no official lunch. I was now thoroughly out of sorts, but my companion wanted very much to stay for at least part of Spender's poetry workshop, so we did.

I had never been to a poetry workshop and couldn't imagine what one would consist of. I had been to plenty of physics workshops and knew only too well what *they* consisted of—six physicists in a room with a blackboard shouting at each other. The lecture room where Spender was to conduct his workshop was full, perhaps thirty people. One probably should not read too much into appearances, but these people—mostly women—looked to me as if they were clinging to poetry as if it were some sort of life raft. If I had had access to Spender's Journals—they came out a few years later—I would have realized that he was very used to all of this. In fact he had been earning his living since his retirement from University College in London a decade earlier by doing lectures and classes for groups like these. I would also have realized that by 1981 he was pretty tired of it, and pretty tired of being an avatar for his now dead friends—Auden, Eliot, C. Day Lewis, and the rest. He had outlived them all but was still under their shadow, especially that of Auden, whom he had first met at Oxford at about the same age and the same time that Oppenheimer had met Dirac.

Spender walked in with a stack of papers—poems written by the workshop members. He gave no opening statement but began reading student poems. I was surprised by how awful they were. Most seemed to be lists—"sky, sex, sea, earth, red, green, blue. . . .” Something like that. Spender gave no clue as to what he thought of them. Every once in a while he would interrupt his reading and seek out the author and ask questions like “Why did you choose red there rather than green? What does red mean to you?” He seemed to be on autopilot.

It is a pity that there are no entries in his Journal for this precise period. But it is clear that he was leading a rich social life at the time: dinner with Jacqueline Onassis one day, the Rothschilds at Mouton a week later—the works. My feeling was that whatever he was thinking had little to do with this workshop. Somehow I was getting increasingly annoyed. It was none of my business, I guess, but I had put in a long day, and I felt that Spender owed us more.

My companion must have sensed that I was about to do something because she began writing furiously in a large notebook that she had brought along. Finally, after one particularly egregious “list,” I raised my hand. Spender looked surprised, but he called on me. “Why was that a poem?” I asked. In reading his Journals years later I saw that this was a question that he had been asked by students several times, and he had never come up with an answer that really satisfied him. In 1935 Auden wrote an introduction for an anthology of poetry for schoolchildren in which he defined poetry as “memorable speech.” That sounds good until one asks, memorable to whom? Doesn’t it matter? If not, why a workshop?

I can’t remember what Spender answered, but I then told him that when I was a student I had heard T. S. Eliot lecture. After the lecture one of the students in the audience asked Eliot what he thought the most beautiful line in the English language was—an insane question, like asking for the largest number. Much to my amazement, Eliot answered without the slightest hesitation, “But look, the morn in russet mantle clad / Walks o’er the dew of yon high eastward hill.” I asked Spender what he thought the most beautiful line in the English language was. He got up from his chair and in a firm hand wrote a line of Auden’s on the blackboard. He looked at it with an expression that I have never forgotten—sadness, wonder, regret, envy. He recited it slowly and then sat back down. There was total silence in the room. I thanked him, and my companion and I left the class.

I had not thought of all of this for many years, but for some reason, perhaps a new biography of Auden, recently it all came back to me—except for the line that Spender wrote on the blackboard. All that I could remember for certain was that it had to do with the moon, somehow the moon. My companion of fifteen years ago is my companion no longer, so I could not ask her. I am a compulsive collector of data from my past, mostly in the form of items that were once useful in tax-preparation. I looked in the envelopes for 1981 and could find no trace of this trip. Then I had an idea—lunatic, lunar, perhaps. I would look through Auden's collected poems and seek out every line having to do with the moon to see if it jogged my memory. One thing that struck me, once I started this task, was that there are surprisingly few references to the moon in these poems. In a collection of 897 pages I would doubt if there are 20. From "Moon Landing" there is "Unsmudged, thank God, my Moon still queens the Heavens as she ebbs and fulls. . . ." Or from "The Age of Anxiety," "Mild, unmilitant, as the moon rose / And reeds rustled. . . ." Or from "Nocturne," "Appearing unannounced the moon / Avoids a mountain's jagged prongs / And sweeps into the open sky / Like one who knows where she belongs." All wonderful lines, but not what I remembered. The closest was, "White hangs the waning moon / A scruple in the sky . . . ," also from "The Age of Anxiety." This still didn't seem right.

Then I got an idea. I would reread Spender's Journals to see if he mentions a line in Auden's poetry that refers to the moon. In the entry for February 6, 1975, this is what I found, "It would not be very difficult to imitate the late Auden. [He had died in 1973.] For in his late poetry there is a rather crotchety persona into whose carpet slippers some ambitious young man with a technique as accomplished could slip. But it would be very difficult to imitate the early Auden—'This lunar beauty / Has no history, / Is complete and early.' . . . " This, I am sure now, is the line that Spender wrote on the blackboard that afternoon in 1981.

Poor Stephen Spender, poor Robert Oppenheimer, each limited, if not relegated, to the category of the merely very good and each inevitably saddened by his knowledge of what was truly superior. As Spender wrote, "Being a minor poet is like being a minor royalty and no one . . . is happy as that." As for Oppenheimer, I recall I. I. Rabi once telling me that "if he had studied the Talmud and Hebrew rather than Sanskrit, he [Oppenheimer] would have been a much greater physicist. I never ran into anyone who was brighter than he was. But to be more original and profound I think you have to be more focused."

As Spender says, W. H. Auden's poetry cannot be imitated, any more than Paul Dirac's physics can be. That is what great poetry and great physics have in common: both are swept along by the tide of unanticipated genius as it rushes past the merely very good.

Shadows

In the preface to his latest book, *In the Shadow of the Bomb,* the physicist and historian of science Sylvan S. Schweber describes its genesis. He tells us that ten years ago his mentor Hans Bethe (pronounced *beta*) asked him to write Bethe's "intellectual biography." This turned out to be a monumental task since Bethe, who is now in his early nineties and still very active, has in the last seven decades made profound contributions to almost all branches of physics. Schweber realized that this would require several volumes to do and that in the end it might not have much interest for nonspecialist readers. Hence he decided to condense it to a more manageable size. Indeed, this relatively short book—186 pages of text and some 70 pages of notes and bibliography—began as a chapter in that book, a chapter that expanded to the point where Schweber decided to make it a book in its own right. While there is much in this monograph—which is a kind of dual profile of Bethe and J. Robert Oppenheimer—that I admire, I am constrained to say that, as a whole, I found it to be something of a disappointment.

Part of the problem is something that Professor Schweber (he is at Brandeis University) explains in the preface. He writes, ". . . I know so much more about Bethe's life than about Oppenheimer's, and Bethe's story continues to unfold." This works against his project in two ways. The obvious one, which I will discuss in more detail, is that his portrait of Oppenheimer does not really come to life. The less obvious one is that the presence of the living Bethe, whom—and with good reason—Schweber admires enormously, inhibits his portrait of Bethe. Take the following. Schweber tells us that in the 1970s Bethe's activities in pure sciences, as opposed to governmental and industrial consulting, were not "outstanding." He notes that part of the reason had to do with "crises within the home." That is the end of it. We are not given a clue as to what this refers. Either one is writing a biography or one is not. If one does not wish to explain something like this, then why mention it at all? Furthermore both the Bethe and Oppenheimer profiles are encumbered with material which, as far as I am concerned, only get in the way of the story. These digressions might be ignorable in a long book, but in a short one they assume almost Everest-like proportions, especially since they take the place of what I, at least, regard as essential. Let me illustrate first in the case of Oppenheimer.

Oppenheimer, who was born in New York in 1904, came from a well-to-do German-Jewish family and was educated at the Ethical Culture High School in New York. This motivates Schweber to make a multi-page and somewhat tedious digression into the ideas of the educator Felix Adler, who founded the school. The question of what any of this had to do with Oppenheimer is not especially clear. What is clear is that Oppenheimer had a great deal of difficulty with his Jewish heritage. Where did this come from? The school? His fellow students? Ethical culture? It must be noted that in the 1920s and later, anti-Semitism in this country was part of the furniture. Schweber quotes from a letter of recommendation that Oppenheimer's great teacher at Harvard, Percy Bridgman, wrote in 1945 to Ernest Rutherford in England, suggesting that Oppenheimer come work with him. But it is the part he does not quote that is the most interesting. In it Bridgman notes that "As appears from his name, Oppenheimer is a Jew, but entirely without the usual qualifications of his race. He is a tall, well set-up young man, with a rather engaging diffidence of manner, and I think you need have no hesitation whatever for any reason of this sort in considering his application." Bridgman was in no way an anti-Semite. He was trying to *help* Oppenheimer. Jews with a stronger emotional center of gravity were able to deal with this sort of attitude, often by finding strength in their heritage. It seems to have torn Oppenheimer apart. One of his oldest friends, the physicist I. I. Rabi, told me that Oppenheimer reminded him of a man he knew who could not decide if he wanted to be a member of the B'nai B'rith or the Knights of Columbus. He added that if Oppenheimer had studied Yiddish rather than Sanskrit, he might have become one of the best physicists who ever lived.

I have often wondered if this insecurity contributed to the indiscretion that Oppenheimer sometimes showed in conversation. I witnessed this firsthand. In the fall of 1957 I had just driven across-country to Princeton to take up a postdoctoral appointment at the Institute for Advanced Study, of which Oppenheimer was the director. I was told that Oppenheimer wanted to see me. Disheveled as I was, I went to his office to find him, as usual, impeccably dressed in one of the bespoken suits he had tailored at Langrocks, the local tailor to the fashionable. He studied me with those remarkable blue eyes and asked, "What is new and firm in physics?" The "and firm" was especially good. Before I could try to answer, the phone rang. I got up to leave, assuming he would want to take the call in private. He motioned me to remain seated and carried out a brief, personal, conversation. When he hung up he said to me, "It was Kitty," a reference to his wife, whom I had never met. "She has been drinking again." That he would say this to me, a perfect stranger, left an impression I have never gotten over. As thoughtless as it was, it was not likely to have any repercussions. But some of Oppenheimer's indiscretions did have repercussions. Professor Schweber provides an example, and I am grateful to have been reminded of the details.

Bernard Peters, who had been born in what was the German city of Posen in 1910, had moved to Munich in 1932 to study electrical engineering. When Hitler came to power, Peters, who was never a Communist, took part, along with some Communists, in anti-Hitler demonstrations. He was arrested and sent to Dachau, from which he miraculously escaped. He managed to bicycle at night to Italy to join the woman whom he later married. After a further odyssey they ended up in the San Francisco bay area where she obtained a research position in the Stanford Medical School while he worked as a longshoreman. He met Oppenheimer socially, and somehow Oppenheimer recognized his abilities and suggested that he come to Berkeley to study physics. After one undergraduate course, Oppenheimer decided that Peters was so gifted that he should be admitted straightaway to the graduate program. He was later invited to come to Los Alamos but decided to remain in Berkeley at the Radiation Laboratory.

The scene now shifts to 1944. By this time Oppenheimer was the director of Los Alamos. For reasons not entirely clear, Peer de Silva, who was in charge of security, brought up with Oppenheimer the names of four of his former students, among them Peters. He asked which of them was most likely to be the most "dangerous." In a moment of monumental indiscretion, Oppenheimer named Peters. In a report that de Silva later filed with the House Committee on Un-American Activities, which was leaked in 1949 to the *Rochester Times-Union* (Peters was then teaching at the University of Rochester), Oppenheimer was alleged to have said that Peters's background "was filled with incidents which indicated his tendency towards direct action"—such as taking part in an anti-fascist demonstration in Germany—and that he was "quite red." In the context of the times such a report could ruin someone's life. A phrase like "quite red" could mean anything.

When Oppenheimer's colleagues, such as Bethe, read this newspaper report they were outraged, not only that it had been leaked but that Oppenheimer could have made such a set of statements, especially when on several occasions he had spoken highly of Peters. Peters finally had a chance to confront Oppenheimer. It must have been a dreadful scene. Oppenheimer could not deny that he had made the statements but insisted that his action had been a "dreadful mistake." He called the University of Rochester and received assurances that Peters's position there was not in jeopardy. He even wrote a letter to the Rochester newspaper, which did more harm than good. While Peters did not lose his job, his travel was restricted. In 1951 he took a position in India, then in Copenhagen. When he died in 1993 he was recognized as one of the most significant contributors to his field—cosmic rays. Without any pretense of being a psychoanalyst, my guess is that the same impulse that made Oppenheimer tell me that his wife was an alcoholic was at work when he told de Silva that Peters was the most dangerous student he had. In both cases, it seems to me, it was Oppenheimer's insecurity that made him go beyond reasonable bounds.

One adjective no one would use to label Bethe is "insecure." He often reminds me of a large, inexorable ocean liner. This is quite remarkable considering all the forces at work that could have fragmented Bethe's psyche. His mother was born Jewish but had become a Lutheran before she married his Protestant father. Bethe, who was born in Strasbourg in 1906, grew up in a non-Jewish home, but one that was divided. His parents divorced. Added to this internal turmoil was the external turmoil of the post–World War I period in Germany. I don't think that Schweber does justice to this. Instead he spends many pages on another, to me, tedious divagation, into Wilhelm von Humboldt, Kant, and *Bildung*—the sort of murky metaphysics that Germans used to be so fond of. During the nearly two years I spent interviewing Bethe for a *New Yorker* profile, I was struck again and again by the echoes of this postwar period in both Bethe's life and the life of Germany. Bethe's recollections of the monetary inflation were as vivid as if it had happened yesterday. He described being sent early in the morning to stores to buy food before the currency devalued to the point where, later in the day, it bordered on the worthless. He was quite persuaded that this could happen in the United States, and he was protecting himself by investing in things like rare postage stamps, which he hoped would retain their value. (Curiously, when some of the German-Jewish refugees came to the United States they brought stamp collections. I have such a collection given to my father by a grateful German refugee he had helped bring here.) If all of this was not enough, in 1933 Bethe was summarily dismissed from his university post because of his Jewish grandparents. Among the many things that struck him was that there was essentially no sympathy from his German colleagues. Why then did Bethe not suffer the same sort of centrifugal psychological forces that fragmented Oppenheimer? I can only make a guess.

It is clear that from the beginning Bethe knew how good he was at the thing he wanted to do, theoretical physics. I would imagine that he would have found the diversions that Oppenheimer invented for himself, such as writing poetry, somewhat absurd or at least beside the point, the point of doing physics.* Bethe is able to take almost any problem in physics and turn it into a sensible subject of research.† Since the source of such problems seems limitless, one can lose oneself in them for a lifetime. This ability was recognized by others very early, so that

*Einstein wrote a good deal of semi-comic doggerel.

†Bethe is, incidentally, very good at mental arithmetic, which most theoretical physicists are not. He can make very rapid and accurate order-of-magnitude estimates of answers to very complicated problems.

Bethe was able rapidly to relocate himself in England. There being no permanent job available there, he soon accepted an offer from Cornell University where he has been ever since, despite numerous attempts to lure him away. I do not think he knew Oppenheimer at all well before the war. They were on different coasts and worked on entirely different problems. Just before the war Oppenheimer and his students presented the first modern theory of what later came to be called "black holes." While this was interesting, it must have seemed to many physicists as effete as the poems in *Hound and Horn*. But these ideas, much expanded, are now in the forefront of contemporary research. Bethe, on the other hand, had immersed himself in nuclear physics. In 1938 he proposed the set of nuclear reactions—the so-called "carbon cycle"—that provides the energy emitted by many stars. He won the Nobel Prize for this in 1967. He had also written a massive review article on nuclear physics which came to be known as "Bethe's bible."

The choice of Oppenheimer as the director of the Los Alamos laboratory struck most of his colleagues as almost incomprehensible. In the first place Oppenheimer was not a nuclear physicist. He was not even an experimental physicist. His early forays into experimental physics had been disastrous. He was not an engineer. He was notorious for getting arithmetic factors wrong. He had never run a large engineering project. To add to all of this, he came with a burden of left-wing associations. His brother had been a member of the Communist party, and his wife had been married to a Communist. Some of his students had had flirtations with the party, and Oppenheimer himself, while certainly never a member of the party, had had associations with Communist front organizations. It is unlikely that he could have been cleared to work on radar, which in the beginning of the war was the really important super-secret military project. Nonetheless General Leslie Groves, who was in charge of the nuclear weapons program, selected him. He must have had some instinct, and in this case his instinct was absolutely right.

One cannot be certain that without Oppenheimer the atomic bomb could not have been built, when it was built. Sooner or later it would have been—perhaps not in this country. There is no example I can think of where once something like this is shown to be possible in principle, it has not been done. Human beings do not seem to have the gift of restraint when it comes to creating weapons of mass destruction. Whether their *use* can be restrained is an open question. What made Oppenheimer so good? One thing was his essentially instantaneous comprehension of every facet of the science and engineering. One must read the technical accounts of this project to understand what this comprehension meant. Everything was new. To take a graphic example: plutonium is not found naturally. It had to be manufactured in reactors that were also new. Its properties had to be studied initially in milligram samples. But then it turned out that reactor-plutonium was contaminated with an unwanted and unavoidable isotope which fissioned spontaneously, potentially making plutonium useless as a bomb. Thus a new method of detonation—implosion—had to be used. This created a whole new engineering science of shaped explosions whose properties—when they could be computed—had to be computed more or less by hand. Electronic computers were still over the horizon.

Oppenheimer was able to understand the problems as they arose and suggest how they could be tackled. He had no difficulty in consulting others and accepting their advice. He also had an élan which is very difficult to capture in words. He had it when I was at the Institute for Advanced Study. He also seemed to have the same instantaneous comprehension, though I was never sure how deep it went. His Delphic comments were often impenetrable. But then nothing much was at stake. If you are trying to design an atomic bomb, your comments have to make sense. During the war he metamorphosed into one of the greatest laboratory directors who ever lived.

I have an ineluctable memory of a train ride I took from Princeton to New York during my stay at the Institute. After I was seated, Oppenheimer got on board and, as I was the only familiar face, he sat down next to me. During the next hour or so we talked. Actually he did most of the talking. One of the things he discussed—I cannot imagine how this came up—was what he referred to as "my case." These were the hearings of the Atomic Energy Commission in 1953 at which Oppenheimer lost his security clearance. What struck me was his apparent detachment when he spoke of it. It was as if it had not happened to *him*. Schweber describes Oppenheimer's defense at the hearings as "apathetic." There is some truth in this. But Schweber attributes this in part to Oppenheimer's need to do "penance," presumably for his role in creating the atomic bomb. I am not sure this is right. I think it was more like disbelief, disbelief that after having performed what he regarded as a patriotic wartime service, he was being tried for things for which he had been cleared several times during the war. I am not sure that at the time he understood the real reason for these hearings, which was to discredit him in such a way that he would lose his public voice. If the Atomic Energy Commission no longer wanted his advice, it could simply have chosen not to renew his consulting contract. But this would have left him as a highly visible opponent of the new reliance of the military—above all the air force—on the nuclear option. Once he was publicly disgraced, he was effectively silenced. I did not know Oppenheimer before the hearings, but people like Bethe and Rabi who did, told me that he was a changed man, a diminished man.

The other thing that stands out in my memory about this conversation on the train was Oppenheimer's telling me that he was thinking of writing a play. I don't know if this was literally true. T. S. Eliot had been at the Institute in the fall of 1948* and had written a play. Maybe that had inspired Oppenheimer, though he told me that he did not much like Eliot's play *The*

*Oppenheimer had been reading Eliot's poetry since his undergraduate days at Harvard.

Cocktail Party. In any event, Oppenheimer told me that his play was going to be called *The Day That Roosevelt Died.* Oppenheimer felt that Roosevelt's death put an end to the possibility of cooperation with the Russians and a possible international control of the bomb. I don't think he appreciated the Russian feelings about the bomb. Truman seems to have had a more realistic intuition about this than Oppenheimer or, for that matter, Roosevelt. The one time Oppenheimer tried to discuss these matters with Truman was a disaster. He began by telling Truman that he, Oppenheimer, had "blood on his hands" for creating the weapon. Truman was furious since he was the one who had ordered its use. He got rid of Oppenheimer as fast as possible and told Dean Acheson, who had arranged this meeting, never to arrange another. We now know from reading the memoirs of people such as Andrei Sakharov that nothing—force aside—would have stopped Stalin from trying to make a bomb. The Russians knew because of the espionage by people such as Klaus Fuchs, well before Hiroshima, that the bomb had been successfully tested and, indeed, how it had been designed. Even Sakharov, who later became one of the outspoken opponents of the testing of nuclear weapons, to say nothing of their use, worked willingly to create them since he felt that his country was in jeopardy.

Bethe's association with nuclear weapons goes back to the summer of 1942 when he was invited by Oppenheimer to join a small group in Berkeley. He made the trip via Chicago. Enrico Fermi's ultimately successful attempt to make the first nuclear reactor was well under way there, and this convinced Bethe that nuclear weapons were a real possibility. He was joined by Edward Teller, and during the rest of the journey they discussed Teller's nascent ideas of trying to make a hydrogen bomb using an atomic bomb—whose design had not yet been made—as the fuse. Bethe was working at the Radiation Laboratory at MIT, which was developing radar. But Oppenheimer persuaded him to come to Los Alamos, which he did in March 1943. Oppenheimer chose Bethe over Teller to head the theoretical division. This caused a great deal of resentment, perhaps even helping to explain Teller's very destructive testimony in Oppenheimer's security hearings. Among Bethe's charges in the division was the very young Richard Feynman. Their monumental daily arguments over technical matters—which Bethe enjoyed immensely—could be heard throughout the division. After the war Feynman came to Cornell with Bethe, and it was there that he did his great work on quantum electrodynamics. Bethe collected a group of brilliant students, which included Freeman Dyson, who never bothered to get a Ph.D. He was made a professor at the Institute for Advanced Study without one.

Both Bethe and Oppenheimer became quite active after the war in trying to educate the general public, as well as the government, about the situation that had been created with the advent of nuclear weapons. Until his security clearance was removed, Oppenheimer was deep inside the government. He had been chairman of the General Advisory Committee of the Atomic Energy Commission (GAC)—the most powerful civilian body involved with nuclear energy—as well as numerous Pentagon committees. He testified from time to time before Congress; when one watches films of these sessions, Oppenheimer looks like a somewhat disdainful professor trying to lecture a not very bright class. Behind the scenes the GAC was wrestling with the question of whether this country should embark on a crash program to build the hydrogen bomb. In October 1949 the GAC recommended against such a program, though a majority of its members, including Oppenheimer, favored continuing research on the possibilities. One of the matters that came up in Oppenheimer's security hearings was a charge that he had dragged his feet when it came to making the hydrogen bomb, and that this had discouraged people like Bethe from getting involved with it, at least initially. In fairness, the entire GAC had, with good reason—no one knew how to make such a bomb—dragged its feet. But the Russians had detonated their first nuclear weapon, Joe 1, in August 1949, and Klaus Fuchs was arrested the next January. This convinced Truman to order a crash program on January 30, 1950. After that, public discussion of the weapon by people inside the government, such as members of the GAC, was forbidden.

Bethe, though he was a consultant at Los Alamos, was not inside the government and was therefore free to speak out against the hydrogen bomb, which he did. He saw it not as a weapon of war but as an instrument of genocide. He did not oppose a research project into the costs and effects of such a weapon. But he made it clear in a letter—quoted by Professor Schweber—written in February 1950 to Norris Bradbury, the director of Los Alamos, that he would not, on his visits, discuss the "super," the name for the design of the hydrogen bomb which was then being discussed and which in fact did not work.

But in the spring of 1951 things changed radically. At this time the Polish-born mathematician Stanislaw Ulam suggested a new design, which was adumbrated by Teller. The essential idea was to use the radiation pressure created by the explosion of a conventional nuclear weapon to rapidly compress and heat a collection of light nuclei that would then fuse, releasing energy. In stars this compression is provided by the force of gravity, and the process is a slow one. It soon became clear that this design might well work and, as is usual in these matters, once this was realized the pressure to build the weapon was insurmountable. Bethe himself succumbed and became what he called the "midwife"—Teller being the "mother" and Ulam the "father." It was successfully tested in the fall of 1952. Bethe has regretted his role in the affair ever since. He feels that the hydrogen bomb should never have been built. He began a campaign to eliminate all nuclear testing. It did not completely succeed, but in July 1963, with Bethe supplying essential technical input about the detection of underground explosions, a treaty was signed with the Soviet Union which banned atmospheric testing. This has had the curious effect that the present generation of people who work on nuclear weapons have never seen a nuclear explosion. While this is certainly good for the environment, one wonders what it means for inhibiting these efforts. No one who has seen an actual nuclear explosion can think the same way about them afterward.*

Oppenheimer died in 1967 at the age of sixty-three. He never wrote an autobiography, though he did grant some interviews with the historian of science Thomas Kuhn, which were filmed. By now the people who knew him well are departing the scene, so we may never have a full biography. What none of the biographies I have read capture is the fascination of the man. Nothing he did seemed straightforward or simple. I have one further recollection of him. A young Austrian physicist whom Oppenheimer admired, and with whom I had worked, was visiting Princeton. I gave a little party and as an afterthought invited Oppenheimer. Remarkably he showed up. Even more remarkable was what he was wearing. It was a jacket with leather elbow patches that looked as if the moths had been at it. I guess he thought that his Langrock-bespoken suits would look out of place. On the other hand, we were all wearing our best. It was Oppenheimer who looked out of place.

*I can testify to this firsthand having seen two above-ground tests in the Nevada desert in the summer of 1957.

Kurt Gödel: The Decidable and the Undecidable

In the fall of 1957, as I have noted, I began my two-year fellowship at the Institute for Advanced Study. Although I had been a mathematics major in college, I had by this time committed myself to a career in theoretical physics, a career that I was just beginning. Nonetheless, one of the residues of my years of studying mathematics was a feeling that I can only describe as awe for the work of Kurt Gödel, who was then a professor at the Institute. In a brief series of papers written in the early 1930s, when Gödel was in his mid-twenties, he transformed forever the way we view mathematical truth. Before his discoveries it was generally assumed that mathematical systems, like geometry or the theory of numbers, rested solidly on a foundation of extremely plausible axioms and definitions—for example, between any two points there is one and only one straight line—which were connected to the theorems—for example, the sum of the interior angles of a triangle is 180 degrees—by a stainless steel webbing of logical argument. What was true was provable. Gödel showed almost the exact opposite. First he showed that in systems complicated enough to include the usual numbers and their properties, there were necessarily propositions that were *undecidable.* They might well be true, but no proof of this could in principle exist within the system. Moreover, among these undecidable propositions was that of the consistency of the axioms themselves! You could never prove that your axioms would not lead to a logical catastrophe. You might find one day that the axioms implied both the truth *and* the falsity of the same proposition. The castle you thought you were living in might turn out to be a house of cards.

I had studied Gödel's discoveries in college, and that is why I held him in such awe. Although he was in some sense my neighbor at the Institute—he had his office in the next building—it would never have occurred to me to try to visit him there. I could not imagine what I would have had to say to him. Moreover he had the reputation of being "reclusive." At the time I was not exactly sure what this meant, except that it meant that he did not welcome casual visitors. That fall, however, Robert Oppenheimer decided that it might be nice to hold what the Radcliffe girls used to call a "jolly up"—a little social where we could all get to meet one another. It was held, as I recall, in the Institute cafeteria, where the usual suspects—professors, visitors, the odd local—had all been rounded up. There in the corner, much to my astonishment, was Gödel. Why he came to this "jolly up" I cannot imagine, but there he was. I recognized him instantly from his photographs. He was exceedingly thin and looked central European in the same sense that Kafka looked central European. Perhaps it was the dark horn-rimmed glasses. (I used to wonder if there was one factory in Austria that supplied them for the entire Austro-Hungarian Empire.)

Gödel was in one corner and I was in another corner. But then something unexpected happened. He started to be introduced around, perhaps by Oppenheimer, and it soon was my turn. When I told him my name, he said, "I knew your father in Vienna." Here was a proposition that was not only decidable but decidedly false. My father was a rabbi in Rochester, New York, and to the best of my knowledge had never set foot in Vienna. I pointed this out politely to Gödel, who then repeated in exactly the same tone of voice as before, "I knew your father in Vienna." I realized that whatever the problem was, it was not going to be solved at a "jolly up." I thanked him, and he moved on to the next guest. I spent the next three days trying to figure out who Gödel had in mind. Then it dawned on me. In the theory of sets—the theory of ensembles of objects, to which Gödel also made monumental contributions—there is a famous theorem called the Schröder-Bernstein theorem. Ernst Schröder and Felix Bernstein were two German mathematicians who proved the theorem independently. Bernstein was some ten years older than Gödel. During the war he emigrated to the United States. Somehow Gödel had decided that I was Bernstein's son. How Vienna came into it I did not understand. Since I never had another chance to speak to Gödel, I never found out.

All of this came back to me when I read a recent biography of Gödel written by the mathematical logician John W. Dawson, Jr., entitled *Logical Dilemmas: The Life and Work of Kurt Gödel.* It is with some reluctance that I use the term "recent biography," since it might imply that there is a string of biographies of which this is the most recent. In fact, since Gödel's death in 1978, apart from brief biographical sketches, there has been no biography. The reasons are not difficult to find. To write a biography of Gödel one must really understand what he did, and I believe this is something that only a professional mathematician or mathematical philosopher can do. Dawson is a professor of mathematics at Pennsylvania State University. In the second place, there is the character of Gödel to deal with. To call him "reclusive" is to trivialize the situation. Gödel was, certainly in the last years of his life and on and off for most of it, a full-blown paranoiac. He granted, as far as I know, almost no interviews. From time to time he would respond to letters sent to him about his life and work, but in many cases these responses were never actually sent and were found only after his death in his *nachlass*—his personal papers—which over a period of two years Professor Dawson catalogued along with the rest of Gödel's unpublished manuscripts. Among the *nachlass* were library slips for every book that Gödel had checked out of any library from the time he had been a student in Vienna in the 1920s. Given all of this, the fact that he appeared at Oppenheimer's "jolly up" in 1957 strikes me as nothing short of a miracle. Professor Dawson has put Gödel's life and work together with understanding and respect. Indeed, one wonders if anyone can write another biography of Gödel.

Kurt Gödel was born in Brno, Moravia, on April 28, 1906, and baptized as a Lutheran, though both his parents were non-Lutheran Christians. I mention this only because later in his career people seem to have assumed that Gödel was Jewish. Professor Dawson observes that in his *Autobiography* Bertrand Russell recalls that after the war, when he was staying in Princeton, he used to go to Einstein's house once a week for discussions with Einstein, Gödel, and the physicist Wolfgang Pauli, which he said were "somewhat disappointing." He noted that "although all three of them were Jews [sic] and exiles [Pauli, who was half Jewish, was a Swiss citizen and hardly an "exile"] and, in intention, cosmopolitan . . . they all had a German bias towards metaphysics. . . ." In 1939, not long before he emigrated to the United States, Gödel was beaten up in Vienna by some Nazi street thugs who had taken him for a Jew.

It would appear that there was no indication in the rest of the Gödel family of any special genius in mathematics. (This was, by the way, also true in Einstein's family, to say nothing of Kepler's.) Gödel's father was a rather successful textile merchant, but his mother, on whom Gödel doted, actually raised Kurt and his older brother Rudolf, who later studied medicine. As one might imagine, Gödel was an excellent student. I cannot tell if he displayed any of feats of mathematical precocity that one often finds with mathematically gifted children. But even as a young student he already began exhibiting the kind of isolation and withdrawal that was so characteristic of his adult life. He always had an interest in foreign languages, which began in high school. His *nachlass* contains notebooks on foreign languages, and his library had various foreign-language dictionaries and grammars. He also learned shorthand.

In 1924 Gödel matriculated at the University of Vienna. He intended to study physics, but after a couple of years he gravitated toward mathematics, in part due to a gifted mathematics teacher, Hans Hahn. By the time Gödel met him, Hahn's interests had concentrated on the foundations of mathematics and its philosophy. He was one of the founding members of what became known as the "Vienna Circle," the group of brilliant scientists and philosophers who took it unto themselves to rid science of what they considered to be "metaphysics." The Circle met at various coffeehouses in Vienna. When I studied there in the early 1960s I managed to locate an ancient waiter who said he remembered their meetings vividly. Gödel attended some of these meetings. He apparently said very little, probably because he vehemently disagreed with the Circle's positivistic approach to mathematics. Gödel was then, and remained, a Platonist in mathematics. Mathematical entities have, he thought, a reality that we *discover* and not create. He was, I think, neither surprised nor disappointed to learn from his theorems that there was more to mathematics than what can be generated by logical deduction from axioms.

In 1910 Bertrand Russell and Alfred North Whitehead published a monumental treatise which they called *Principia Mathematica.* I am sure that this title was not accidental. It is part of the title Newton gave to *his* treatise *Philosophiae naturalis principia mathematica,* the one that laid the foundations of the science of mechanics for the next two and a half centuries. What

Russell and Whitehead thought that they had done was to produce a set of axioms and deductive principles from which every true statement in what one usually calls mathematics could be derived. Their *Principia* lasted about twenty years. In between, mathematicians such as David Hilbert, arguably the greatest mathematician of his era, had raised the question of whether one could show that these axioms were actually consistent as well as "complete"—that every true proposition necessarily had a formal proof. Hilbert, to say nothing of Russell and Whitehead, felt strongly that the axioms were consistent and complete. This certainly was the common view. It is not clear what exactly inspired Gödel to question this. There were some hints that Hilbert might be wrong, but it took Gödel to show that he *was* wrong, something that became the content of his Ph.D. thesis. He was in his mid-twenties, about the same age that Einstein was when he showed that Newton's *Principia* was wrong.

As I mentioned before, I don't think this result shocked or disappointed Gödel. He certainly did not feel then, or later, that this showed some kind of limitation of the human mind—something one sometimes hears characterizing Gödel's results. On the contrary, it showed to Gödel that mathematics was not some kind of logic machine. Indeed, some years later the British mathematical logician Alan Turing translated Gödel's results into the language of logic machines—abstract computers. He showed that if one wanted to use such a machine to explore such a net of inferences, there were necessarily some propositions for which one could not be sure that such a procedure would ever terminate. For these propositions the machine would go on grinding forever without letting the user know if the proposition was ultimately true or false. This became known as the "halting problem." It is also ironic that some commentators—whom I find to be totally befuddled—think there is some sort of relation between Gödel's incompleteness theorems and Heisenberg's uncertainty principles in quantum mechanics. It is ironic because

Gödel, like his good friend Einstein, did not believe in quantum mechanics.

Not long after finishing his thesis, Gödel showed that among the undecidable propositions was the consistency of the system itself. This was a surprise to Gödel and everyone else. It also introduced a completely new way of looking at such systems. On the face of it the statement that a mathematical system is inconsistent sounds like a statement *about* the system. It looks as if we are necessarily standing outside the system in order to study one of its properties—its consistency or inconsistency. Gödel's genius was to show how to encode such an apparently meta-systemic question within the system itself. He devised a method of attaching numbers—"Gödel numbers"—to each formula and hence to the strings of formulae that comprised any logical deduction. The proof that the system was consistent then came down to an arithmetic statement. If no such arithmetic statement was possible, one could not decide if the system itself was consistent. That is what Gödel showed.

As one might imagine, the reaction to these results was mixed. Hilbert was very disturbed by them but was too good a mathematician not to come to realize that Gödel was right. John von Neumann, who was about Gödel's age and who was one of the most brilliant mathematicians who ever lived, was well under way toward his own proof of the undecidability of consistency when he found that Gödel had beaten him to it. Not long afterward, von Neumann emigrated to the United States where he became one of the first professors at the Institute for Advanced Study. His admiration for Gödel was limitless, something that turned out to be very fortunate when it came time for Gödel himself to emigrate. Von Neumann and the other professors at the Institute were able to find Gödel a home there. In 1933 Gödel made his first visit to the Institute, where he delivered a course of lectures. But it was upon his return to Vienna in 1934 that he had the first of his nervous breakdowns.

Any biographer of Gödel must confront this aspect of his life. As I mentioned earlier, I think Professor Dawson deals with these matters with sensitivity and compassion. Let us start at the end and reconstruct how things got that way. When Gödel died he weighed sixty-five pounds! The cause of death was given on his death certificate as "malnutrition and inanation" resulting from "personality disturbance." Could this have been avoided? There was always a mixture in Gödel of physical and psychological disorder, the latter being, in one degree or another, paranoia. The pattern was set right from the start. Before he entered the Pukersdorf Sanitorium near Vienna in the early 1930s, Gödel had had an inflammation of the jawbone. This, it seems was due to a bad tooth which Gödel accused his dentist of infecting, perhaps deliberately. For the rest of his life Gödel was sure that doctors were conspiring against him. As his real physical problems accrued, this prevented adequate treatment of them. In 1934 he was treated by a Nobel Prize–winning psychiatrist named Julius Wagner-Jauregg. Wagner-Jauregg concluded that Gödel's breakdown was a consequence of overwork, something which a brief stay in a spa would readily deal with. Considering what Gödel had just accomplished, this seemed plausible enough, but a year later Gödel was back in the sanitorium suffering from depression. Despite this he managed to make a second trip to the United States. This was also characteristic. Gödel always seemed to recover and upon recovery was as brilliant and, at least as far as mathematics went, as lucid as before. One of the things that surely saved him was the relationship with the woman who eventually became his wife.

Her name was Adele Thusnelda Porkert. Her father lived across the street from the Gödels, which was how they met. She was six years older than Gödel and in fact married—unhappily. She professed to have been a ballet dancer, but at the time of their meeting she was employed as a dancer in a Viennese nightclub, Der Nachtfalter. What this meant in reality is unclear, but as far as Gödel's family was concerned—in this case his brother and his mother, as his father had died in 1929—it meant that she was little better than a prostitute. They vehemently opposed Gödel's increasing interest. By the mid-1930s Gödel and the now divorced Adele were traveling together. In 1938 they married. Professor Dawson's book contains a wedding picture. Adele is blonde and rather pretty while Gödel appears almost sleek—dark horn-rimmed glasses and all. Two years later the economist Oskar Morgenstern, who had known Gödel in Vienna (he was a member of the Vienna Circle and had emigrated to the United States where he had become a professor at Princeton) met Adele after the Gödels had themselves emigrated. He confided in his diary that she was a "Viennese washerwoman type: garrulous, uncultured, [and] strong-willed . . ."—someone who would never fit into Princeton society. Be that as it may, she seemed to be able to deal with Gödel's paranoia. She tasted his food to make sure that it had not been poisoned and was able to listen to his complaint that the refrigerator was a dangerous source of poison gas. Gödel had constant stomach problems, and there were very few things he was willing to eat. Even Morgenstern granted that she had "probably saved" Gödel's life. She outlived Gödel by three years, but during the last years of his life she was so burdened by her own mental and physical problems that neither she nor anyone else could stop Gödel's decline.

On March 12, 1938, German troops entered Austria and were greeted with joy and open arms by the population. The Nazification of Austria was complete. Gödel's reaction to all of this is difficult to gauge. Insofar as he said anything about it, it was always oddly detached. Whether this reflected how he really felt or was a form of self-protection is hard to disentangle. The Nazis who then ran the educational system in Austria could not decide about him either. One thing was held against him: his thesis had been supervised by a Jew, Hans Hahn. On the other hand, he seemed so apolitical that none of the authorities had any clear idea of where he stood. Meanwhile von Neumann and others were busy trying to get him a nonquota visa to enter the United

States. Without such a visa the German authorities would not allow him to leave Austria. It was during this period that Gödel was beaten up on the street and emigration became imperative. Finally, in January 1940, the Gödels were able to leave via Berlin. They had to take the trans-Siberian railroad and then go to Japan, where they left by ship from Yokahama to San Francisco.

One of the curious things about Gödel's relation to the Institute was that, despite the fact that he was unarguably one of the greatest mathematicians of this century, he did not have a professorship there until 1953. Until 1946 his membership was renewed annually, and for the next seven years he had some sort of permanent but nonprofessorial appointment. He seems to have run into the sort of viper's nest of intrigue that characterizes most academic institutions. It is also curious that Princeton University did not confer an honorary doctorate on him until 1975, well after he had received all sorts of other honors elsewhere. By this time he didn't care and did not attend the award ceremony. Equally curious about Gödel's stay at Princeton was his relationship with Einstein.

It is hard to imagine two people less alike than Gödel and Einstein. Whatever revisionist notions of Einstein's character have been dredged up in recent years, no one has accused him of paranoia. There radiated from him a supreme sense of self-confidence and serenity. Moreover Einstein never had much interest in pure mathematics as such and even less interest in academic philosophy. For him these were simply tools useful for unlocking the secrets of the "Old One"—Einstein's playful reference to God. Physically he and Gödel were also totally different. The impression that Einstein made on many people was of a physically very powerful man. C. P. Snow, who visited him in the late 1930s, thought he looked like a retired footballer. My colleague T. D. Lee had the opportunity to discuss some physics with Einstein in the 1950s and was struck by the size and strength of Einstein's hands. Gödel, on the other hand, looked as if he would blow away in a windstorm. Nonetheless the two men became very close. Einstein, along with Morgenstern and to a lesser extent von Neumann, took it upon himself to look after Gödel. Professor Dawson suggests that one of the things that attracted Einstein to Gödel was Gödel's capacity of taking some apparently outrageous position and defending it with an intricate logical argument. It is possible that, in some sense, Einstein even enjoyed Gödel's "craziness."

Two anecdotes suggest this. Both are in Professor Dawson's book. The first one I heard myself from, so to speak, the horse's mouth—in this case Ernst Straus, who was Einstein's last mathematical assistant. He told it on the occasion of the Einstein Centennial celebration at Princeton. He recalled that just after the presidential election of 1952, Einstein came into Straus's office to announce that now Gödel had gone completely crazy. Given everything, Straus was hard put to understand what this could possibly mean. Einstein explained, "Gödel voted for Eisenhower!" What Straus did not describe were the circumstances that made it possible for Gödel to vote at all. In December 1947 Gödel went for his citizenship hearings in Trenton, New Jersey. He was accompanied by Einstein and Morgenstern. The difficulty was that Gödel had detected a logical flaw in the Constitution and was quite capable of refusing to swear allegiance to a country which had a logically inconsistent constitution. Morgenstern drove the three of them to Trenton, and on the way Einstein tried to distract Gödel by telling all sorts of anecdotes. It was like trying to stop a train with a Q-Tip. Fortunately the presiding judge was a man named Philip Forman, who had sworn Einstein in as a citizen a few years earlier. The minute Gödel began his disquisition, the judge made it clear that he was not interested. Gödel later described the judge as being an "especially sympathetic person." He also gave Gödel a one-hour lecture "on past and present circumstances in the United States . . . one went home with the impression that American citizenship, in contrast to most others, really meant something."

Einstein was also impressed that Gödel made a significant

discovery in the theory of relativity, something that the textbooks now refer to as the "Gödel universe." Gödel claimed that he was led to his discovery by reading Immanuel Kant on the nature of time. Whatever the source, Gödel had found a new solution to Einstein's cosmological equations in which the universe rotates and makes it possible to travel backward in time. It is not clear that this universe has any connection with ours, but Gödel predicted that there would be a preferred sense of rotation for the galaxies.

A few years ago I interviewed the Princeton physicist John Wheeler, and he told me the following story. In the 1970s Wheeler and two junior colleagues were working on a book on gravitation. They were at the Institute and decided they would pay a little visit to Gödel. It was a pleasant spring day, but they found Gödel in his office wearing an overcoat with the heater on. This was a period during which Gödel constantly complained of being cold. Gödel wanted to know if in the course of their work they had found any evidence of this preferred rotation. He was disappointed to learn that they had not even considered the possibility. But, Wheeler discovered, Gödel was in the process of trying to find his own evidence. He had taken a standard atlas of the stars and was actually measuring angles with a ruler and making statistics. Then Wheeler said to me, "Incidentally, I ran into a man at the Institute a couple of years ago who was working on a biography of Gödel. He had gone through the papers of Gödel, and here were these pages after pages after pages of those numbers. It took him a long time to figure out what they were. Of course they were just this statistical work." And, of course, the man was Professor Dawson.

Professor Dawson ends his book by wondering if anyone will be able to make some sort of theatrical drama out of Gödel's life, the way in which such a drama was made out of the life of Alan Turing. Turing's life *was* a drama. He did brilliant work, including helping to crack the German Enigma code during World War II. He was a homosexual who was convicted under the Gross Indecency Act and later committed suicide by eating a poisoned apple. If there is a drama in Gödel's life it lies in the narrow path he threaded between genius and insanity. He suffered greatly, but he left an intellectual legacy that we can treasure forever.

Giants and Dwarfs

In August 1957 Professor Robert Merton of Columbia University delivered the presidential address before the annual meeting of the American Sociological Association. Merton's specialty was, and is, the sociology of science. Among his several contributions to that discipline is the study of what Merton came to call the Matthew Effect in science. St. Matthew, it may be recalled, remarked that, "For unto every one that hath shall be given, and he shall have abundance; but from him that hath not, shall be taken away even that which he hath." Merton concluded that it was this very principle that governed the reward system in science—prizes, government contracts, and the rest. Having observed the system in operation for some time now, I am entirely inclined to agree with him. "For unto everyone who hath a Nobel Prize shall be given, etc. . . ."

In any event, somewhere around the middle of his presidential address—which the interested reader can find reprinted in full in Merton's collection of articles entitled *The Sociology of Sci-*

ence—there is what seems at first sight to be a fairly innocuous-looking paragraph. Merton is discussing "humility" in scientists which, like the quetzal, is oft spoken of but rarely seen. He writes, "The value of humility takes diverse expression. One form is the practice of acknowledging the heavy indebtedness to the legacy of knowledge bequeathed by predecessors. This kind of humility is best expressed in the epigram Newton made his own: 'If I have seen farther, it is by standing on the shoulders of giants'—this, incidentally, in a letter to his [contemporary and nemesis Robert] Hooke, who was then challenging Newton's priority in the theory of colors."

At the end of this essay I will come back to discuss whether Newton's letter to Hooke was in fact an expression of "humility." But first let me describe what took place after Merton sent a copy of his address to his Harvard colleague, the historian Bernard Bailyn. On November 8, nearly three months later, Merton got Bailyn's reaction—the mills of academe, when they grind at all, do so glacially. Bailyn wrote, ". . . The paper rings all sorts of bells, as you can see: many thanks. By the way [what a "by the way" this turned out to be], I haven't read the Koyré article you cite in ftnote 34 ["ftnote 34" is to the noted historian of science Alexandre Koyré, "An Unpublished Letter of Robert Hooke to Isaac Newton," *Isis,* December 1952]; maybe he goes over the history of the epigram you mention re Newton; but that saying appears to have a rather impressive antiquity. I came upon it twice, in Gilson and in Lavisse, as a remark of Bernard of Chartes in the early 12th century. But Thales [the sixth-century B.C. Greek philosopher] probably said the same thing, only vaguely remembering where he had got it from. . . ." The letter is signed "Bud."

Nearly two months had gone by when on December 30 Bailyn received a reply beginning "Dear Bud." The delay may be understandable. When it was published in a second edition in 1985 by Harcourt Brace Jovanovich, under the title *On the Shoulders of Giants,* the letter from Merton to Bailyn came to 277 pages! It is one of the most joyous romps of scholarship ever recorded. It is unique; nothing quite like it has been seen before or since.

It is not accidental that Merton's book is subtitled "A Shandean Postscript." Merton is a disciple, indeed a "lifelong addict" of *The Life and Opinions of Tristam Shandy, Gentleman.* Readers of the same will recall that Laurence Sterne, its author, raised divagation to an art form. One's mind is jiggled around—divagates—like one of those pollen grains suspended in water that the nineteenth-century Scottish botanist Robert Brown observed through a microscope as they were carrying out "Brownian motion." Even compared to his master, Merton is no slouch. Consider, to take a random example, the following footnote, which is itself a footnote to a divagation that has to do with the various entries in Roget's Thesaurus under the word "dwarf." Among the entries is "Tom Thumb." It sets Merton's wheels spinning. Here is his footnote to a footnote found on page 107 of his epistle. I would do it an injustice if I did not give it in its entirety. Here goes.

"Tom Thumb was not, as some people think, the creation of Barnum, even though the London *Daily Chronicle* of February 6, 1907, might lead you to think so. The *Chronicle* is thoroughly misleading when it says, ' "Tom Thumb" is a name generally given by showmen to lilliputians [and that] the first holder of this "title" was Charles Stratton, who was brought to London by Barnum.' True enough, but the *Chronicle* might have done better to supply more of the historical antecedents of Barnum's coup. The English Tom was actually of Scandinavian descent, as can be inferred from his close connection with the mystic Little Thumb or Tom-a-lyn, Thaumlin, Tamlane, and, not least, Tommelfinger. As early as 1621, R. Johnson was writing *The History of Tom Thumbe,* followed less than a decade later by an anonymous piece entitled *Tom Thumbe. His Life and Death.* Sometime between 1630 and the present, Tom dropped the final 'e' in Thumbe; it is not clear just when. At times the amputation was even more drastic. Needham, for example, writing in 1661 [How in God's name does Merton know any of this? But some, probably all, conjuring tricks are better left unexplained], went so far as to make reference to 'Tom Thums.' And the anonymous *An exact Survey of the affaires of the United Netherlands* [this is too much!] rang in another change in its allusion to 'Tom Thombs.' But by 1700, B.E., who can be no further identified, pretty well settled the whole matter in his *A New Dictionary of the Terms ancient and modern of the Canting Crew,* saying straight-out, and with a certainty that was persuasive enough to carry force right down to our own day, '*Tom-Thumb,* a Dwarf.' (The hyphen fell of its own weight in a few years.)"

Until I read *On the Shoulders of Giants*—"OTSOG," as it is known to the cognoscenti—I had always felt that Gibbon was the nonpareil practitioner of the divagatory footnote. Indeed, in reading *The Decline and Fall* I was often tempted to skip the text and *just* read the footnotes. Who could skip a footnote like the one found on page 269 of the first volume: "The use of *braccae,* breeches or trousers, was still considered in Italy as a Gallic and barbarian fashion. The Romans, however, had made great advances towards it. To encircle the legs and thighs with *fasciae,* or bands, was understood at the time of Pompey and Horace, to be a proof of ill health or effeminacy. In the age of Trajan the custom was confined to the rich and luxurious. It gradually was adopted by the meanest of the people. See a very curious note of Casuabon, ad. Sueton in August.c.82." How Gibbon knows all of this I can't imagine either. Speaking of footnotes, the whole of Nabokov's *Pale Fire* is a footnote for which he should have been given the Nobel Prize for Literature. But I digress.

Speaking of dwarfs, on the other hand, there is an important question that Merton raises, indeed, one that I confess had never occurred to me. Let us suppose the curious arrangement of dwarfs standing on the shoulders of giants was actually achieved, how then would the little fellows maintain themselves once they got there? (My colleague in physics Sidney Coleman, a professor at Harvard, once explained to me that, "If I have seen as far as I have, it is because I have looked over the shoulders of dwarfs," but that again is another matter.) Merton notes that there are schools of thought about dwarfs standing on shoulders. (Among academics there are always at *least* two schools of thought, usually three.) There is what I would call the "head-in-the-sand" school. Its members simply ignore the problem. Merton notes that the French historian Ernest Lavisse, cited in Bailyn's aforementioned letter, is a leading member. In the third volume of his *Histoire de France* he simply refers to the dwarfs as being *hissès*—hoisted—onto the massive shoulders, leaving it up to the tiny homunculi to figure out what to do next. The other two schools of thought identified by Merton might well be called the "sitters" and the "standers." Etienne Gilson, the contemporary French historian who wrote interminably about the Middle Ages, appears to belong to the former. He refers to the dwarfs as being *assis,* like so many passengers in the Metro. This group receives considerable visual support from the stained-glass windows at Chartes. In one of them, St. Matthew—Merton's muse for his eponymous principle—is seated quite comfortably on the shoulders of Isaiah. The fact that they are both about the same size is a cavil. On the other hand, Newton, who started all of this, was clearly a "stander." In no terms uncertain, he *stood* on the shoulders of giants. Merton has discovered that on the Prince's Portal of the Hamburg Cathedral, the statuary shows the apostles standing on the shoulders of the prophets. What this has to do with the subject at hand I am not quite sure. On the other hand, I am no longer sure what the subject at hand actually is.

This leaves us with only two unanswered questions. Who said it first, and what did Newton really mean? Merton addresses the first question but not the second. It would seem that the twelfth-century philosopher Bernard de Chartes is the guilty party. If one is to trust John of Salisbury's rendering of his immediate predecessor, Bernard, whose principal work was his attempt to reconcile Plato and Aristotle—how better to spend a lifetime—remarked apropos, "We are like dwarfs . . . upon the shoulders of giants, and so able to see farther than the ancients." The ellipses, which Merton renders as seven asterisks—*******—represent the uncertainty as to how Bernard actually wanted the dwarfs to be placed. Various translators have placed them variously: seated, standing, or somewhere in between. Of the Russian dancing bear it was said that one should not criticize the quality of the dance but rather be grateful that it danced at all. Of the dwarfs one should perhaps not be overly concerned by their performance once aloft but simply be admiring of the feat of levitation itself.

That having been settled, we are now ready to deal with Newton. What was he up to? To understand this better one must know a little more about Hooke, the recipient of Newton's letter—if indeed, and Merton raises some doubts, he ever did actually receive it. Robert Hooke was born at Freshwater on the Isle of Wight in 1635, making him seven years Newton's senior. Like Newton he was so sickly as a child that he was not expected to survive. But like Newton he lived to a ripe old age. (His dates were July 18, 1635 to March 3, 1703 while Newton's were December 25, 1642 to March 20, 1727.) Also like Newton, he was as a child very gifted in constructing mechanical gadgets of all sorts. Newton went to Cambridge and Hooke went to Oxford, from which, it appears, he never actually took his degree. He did, however, become a member of the brilliant group of scientifically inclined intellectuals who founded the Royal Society. He was interested in everything and buzzed around from field to field making significant contributions whenever he chose to light long enough. He was a scientific genius, but he had the misfortune to share the stage with Newton, perhaps the greatest scientific genius who ever lived. Hooke also had the misfortune to have been, to a certain extent, physically deformed. One contemporary description noted, "As to his person he was but despicable, being very crooked, 'tho I have heard from himself, and others, that he was strait till about 16 years of age, when he first grew awry, by frequent practicing, turning with a Turn-Lath, and the like incurvating exercises, being but of a thin weak habit of Body, which increas'd as he grew older, so as to be very remarkable at last: This made him but of low Stature, tho' by this limbs he should have been moderately tall. . . ." In short, Hooke was a dwarf, or nearly so. In his *Brief Lives* John Aubrey, who was nothing if not pithy, describes Hooke as "but of middling stature, something crooked, pale faced, and his face but little belowe, but his head is large; his eie full and popping, and not quick; a gray eie." I do not pretend to understand what Aubrey meant by all of this, but it certainly does not sound good.

In the seventeenth century, as now, certain scientific problems were in the air, such as the nature of planetary motion, or light, which attracted the best scientific minds. So it is not unexpected to find Newton and Hooke each working in his own way on very similar problems. Hooke was a brilliant inventor of scientific ideas, many of which he did not bother to work out. He suggested, for example, that planetary motion was governed by a gravitational force that diminished with the reciprocal of the square of the distance from the planets to the sun. But he seemed to think that it was beneath his dignity to try to fit the exact orbital shapes. That was for the drones. Newton, on the other hand, felt, as all the great scientists do, that the devil was in the details. For working out his law of gravitation, Newton invented the differential and integral calculi and spent years getting the orbits right. Hooke's cavalier attitude about details drove Newton almost berserk. For example, he wrote to Edmond Halley—he of the comet—that Hooke "has done nothing & yet written in such a way as if he knew & had sufficiently hinted all but what remained to be determined by the drudgery of calculations & observations, excusing himself from that labour by reason of his other business: whereas he should rather have excused himself by reason of his inability. . . ." The feelings were perfectly mutual. In his diary Hooke has an entry in which Newton is described as the "veryest knave in all the Ho. . . ."

This raises the obvious question of why Newton wrote his letter, so full of apparent modesty and respect, to Hooke. To me, this was pretty well answered in Frank Manual's fascinating psychological study, *A Portrait of Isaac Newton.* Some professional historians have taken Manual to task, but I find what he says pretty convincing. The letter, Manual points out, was written not long after Hooke had accused Newton of taking his theory of light from Hooke's *Micrographia.* This, as one might imagine, did not sit well with Newton, who in turn accused Hooke of taking most of what *he* had done from Descartes. After noting this in the oiliest of terms, Newton then goes on, in the now celebrated letter, to describe, again in terms of excessive flattery, the few additions that Hooke had made to Descartes's work. Manual writes, "Following hard upon such dithyrambs, the image of a dwarf on the shoulders of a giant sounds like an abrupt shift in tone, until one realizes that there is something devious about Newton's applying this hackneyed simile to his relationship with Hooke. On the face of it Newton appears to be calling Hooke a giant and suggesting that he is a mere dwarf by comparison; but this hyperbole is, after all, addressed to a 'crooked' little man, and there is an undertone of contempt here, conscious or unconscious, as in calling a fat man 'skinny' and thus underscoring his obesity." In short, and this is the point, it was not so much that Newton wanted to stand on Hooke's shoulders, but rather to step on his head. The shoulders of giants indeed!

The German Atomic Bomb

Anyone who has studied the attempt to develop nuclear energy in Germany during World War II immediately confronts three questions:

1. Were the Germans trying to make a nuclear weapon?

2. If so, how close did they come?

3. If they had succeeded, would they have turned this weapon over to Hitler?

All three of these questions have been answered in very different ways, but before I give my answers, and the alternatives, it is important to clarify what one means by "the Germans." *Which* Germans? To understand the issue, let us review a little of the history. In December 1938 the German physical chemists Otto Hahn and Fritz Strassmann made the first observation of the fissioning of atomic nuclei. They bombarded uranium with slow neutrons and noted that one of the end products seemed to be the production of the relatively light nucleus barium. A month later the Austrian-born Jewish physicists Lise Meitner and her nephew Otto Frisch—who had taken refuge in Sweden and Denmark, respectively, after fleeing Germany—made the correct analysis of the Hahn-Strassmann observation, namely, that the uranium nucleus had been split—barium being one product—with the release of energy. Clearly Meitner, who had been a colleague of Hahn's in Berlin, and Frisch, who had been in Hamburg, should not, when answering these questions, be counted among "the Germans." Nor should Rudolph Peierls, who had emigrated to England from Germany. It was the 1940 memorandum of Frisch (by that time also in England) and Peierls that persuaded the Allies that an atomic bomb was a real possibility. Frisch and Peierls said that if you could separate less than a kilogram—sixty kilograms is closer to the correct value—of the rare uranium isotope U-235 from the common isotope U-238, you could generate an explosive nuclear reaction. Before this calculation it was thought that tons would be required, an amount so massive that scientists such as Niels Bohr had decided that nuclear weapons were a practical impossibility.

One also cannot count among "the Germans" such physicists as the nobelist James Franck, or Hans Bethe or, needless to say, Albert Einstein, who were driven from Germany in the early 1930s. Nor can you count non-Germans such as John von Neumann and Eugene Wigner, both Hungarian Jews, who were beginning their careers in the German university system before they were forced to emigrate. And there were Italians such as Emilio Segré and Enrico Fermi who were forced to leave Italy because of the racial laws, which were a German infection.

But even inside Germany there were German scientists one cannot include among "the Germans." Strassmann himself, as an avid anti-Nazi, was deprived of his livelihood during the war. And then there was the remarkable Max von Laue, also a Nobelist, who publicly refused to deny Einstein credit for his relativity theory, something that most of the other German physicists were all too ready to do. It is a miracle that Laue did not end up in a concentration camp. There is also the notable case of Gustav

Hertz. Hertz, who had Jewish ancestry, shared the 1925 Nobel Prize for Physics with James Franck. His students and colleagues had such admiration for him that when he lost his university position they hid him in the Siemens industrial laboratories in Berlin during the entire war. He is also not one of "the Germans."

There were also Germans whose racial pedigrees were acceptable but who did not have the sort of academic prestige required for them to be taken entirely seriously. A case in point is Manfred von Ardenne. He was an inventor and a scientific entrepreneur of significant talent, though he had no academic qualifications in physics. He managed to persuade the German Post Office to sponsor work in nuclear physics, which was carried out on his estate in Berlin. In early 1941 one of his associates, Fritz Houtermans, who had had the unusual distinction of having been jailed by both the Soviet secret police and the Gestapo, made the observation that what came to be called plutonium was an even better nuclear explosive than uranium. The same discovery had been made independently by the more conventional German physicist C. F. von Weizsacker and by the American physicist Louis Turner a few months earlier. Von Ardenne's group also made significant progress in the separation of the uranium isotopes. While what they had done was more or less ignored by the German scientific establishment, it was important enough so that when the Russians occupied Berlin they shipped von Ardenne, his equipment, and his colleagues east, where they helped make the first Soviet bomb. These people too should probably not be counted among "the Germans." Who, then, is left?

In the fall of 1939 German Army Ordnance decided that it was imperative to sponsor a program to study nuclear fission for use as a possible weapon. They took over the Kaiser Wilhelm Institute for Physics in Berlin and began drafting physicists and chemists to work on the project. The most important "draftee" was Werner Heisenberg, who was certainly one of the greatest physicists of this century. It was Heisenberg who had the first truly quantum mechanical mind. He was to the quantum theory what Einstein was to relativity. His prestige in Germany was enormous. In addition to Heisenberg, some sixty scientists from various institutions eventually joined the project. They came to call themselves the *Uranverein*—the Uranium Club. Clearly these are "the Germans" one wants to consider. But there was also an industrial infrastructure. To take one notorious example, the Degussa Company had taken over a metallurgical firm called Auer since its previous owners were not Aryan enough. Degussa-Auer used slave labor—some two thousand women from the concentration camp at Sachsenhausen—to produce uranium oxide for the *Uranverein.* These too should be counted as part of "the Germans." (Degussa, incidentally, is still in business. It has a web site with a smiling face. One of its activities has been supplying Iraq with nuclear material.) We may now return to the three questions.

First, was the *Uranverein* trying to make a nuclear weapon? It is hard to see how there can be much of an argument about this, at least until early 1942 when the army withdrew from the project, which was then funded by the Reich Research Council. At that point most of the effort was on trying to make a functioning nuclear reactor. This enabled the Germans and their spokesmen to claim that the project was really about the peaceful use of nuclear energy. In 1958 the journalist Robert Jungk published *Brighter Than a Thousand Suns,* in which he stated that "It seems paradoxical that German nuclear physicists, living under a saber-rattling dictatorship, obeyed the voice of conscience and attempted to prevent the construction of atomic bombs, while their professional colleagues in the democracies, who had no fear, with very few exceptions concentrated their whole energies on the production of the new weapon." But, as Jungk himself came to realize, though many years later, the discovery of plutonium turns a reactor into a weapon. Reactors are used to make plutonium, a fact that the *Uranverein* was very explicit about when they made appeals to the government for funding. There was also a smaller program to try to design some sort of exploding reactor—like a miniature Chernobyl—which could have spread radioactive material over a large area. So to the first question my answer is "yes."

Then how close were they to making a bomb? To this I would answer "not very." A variety of explanations have been offered, including the fact that Germany was being bombed and in any event did not have the industrial capacity; that *Uranverein* was not trying and even attempted to sabotage the project; and, my own favorite, incompetence. One must keep in mind that in early December 1942 Enrico Fermi, with an infrastructure that was certainly no greater than that available to the *Uranverein,* succeeded in making the first functioning nuclear reactor in an abandoned squash court at the University of Chicago. This was something the Germans never achieved. The difference, as far as I am concerned, is that the American program had Fermi while the Germans had Heisenberg. Heisenberg was simply not a good engineer, something that his ego prevented him from acknowledging. If the Germans, who started first, had been able to create a self-sustaining chain reaction, I think the whole project would have taken on a different sense of urgency. While they would never have had the industrial capacity to create an Oak Ridge, where uranium was separated in the United States, they might have been able to make a Dimona, the Israeli plutonium-producing reactor which probably produces enough material for a few bombs a year.

This brings me to the last question: would they have turned the bomb over to Hitler? Here our own experience may be relevant. Once the Manhattan Project to build the bomb was launched in December 1941, it became a program of the United States Army under the command of Major General Leslie Groves. Men were drafted and sent as soldiers to the laboratories at Los Alamos. In the beginning there was even a discussion of giving the scientists simulated ranks and putting them into uniform. When the bomb was built, the army simply took possession of what it had bought and paid for. A few of the scientists tried to enter into the process of deciding what to do with the bomb by organizing an action group, but they had no say in how it was finally used. Can anyone imagine that it would have been any different in Germany? If the project had shown more signs of succeeding, the army would have kept control of it and the bomb would have become the weapon of choice.

What is remarkable is that now, a half-century after the fact, the activities and intentions of the *Uranverein* still provoke vivid and polemical debate. To take two recent examples. In 1993 the journalist Thomas Powers published his book *Heisenberg's War,* in which he tried to argue that Heisenberg deliberately sabotaged the German nuclear project by withholding knowledge about the bomb and even attempting to pass information to the Allies. Powers's book unleashed a hailstorm of protest, the most recent example of which is *Heisenberg and the Nazi Atomic Bomb Project* by the Pennsylvania State historian Paul Rose. The title is a clue to the tone. One may wonder why these matters have not been settled. After all, you are not dealing with the reconstruction of a prehistoric civilization from a few drawings on the wall of a cave. This was a project that created a paper trail of hundreds and hundreds of documents of all kinds. The principals were, at least until a few years ago, all alive and prepared to tell their stories. How then are such radical disagreements possible? I will give two examples, both suggested by Rose's book, examples that will take us deeper into the subject and will show just how difficult it is.

Early in his book Rose quotes what he refers to as a "bizarre letter" purportedly written in the spring of 1970—the letter is undated—from Heisenberg to a woman named Ruth Nanda Anshen. Anshen was the editor and guiding spirit of a series of books—called World Perspectives—by outstanding thinkers, including a book by Heisenberg and one by I. I. Rabi, both of whom were on her board of editors. This is the letter. It is taken directly from Anshen's book *Biography of an Idea.*

Dear Dr. [sic] Anshen:

I have finished reading in your Perspectives in Humanism [sic] series the volume written by Professor Rabi entitled *Science: The Center of Culture.* I should like to review this important volume. However, I must say to you that I shall have to take exception to Dr. Rabi's statement that "such a tremendous undertaking as Oak Ridge, with huge, combined efforts of science, engineering, industry and the Army would have been impossible in bomb-ridden Germany.["]

Dr. Hahn, Dr. von Laue and I falsified the mathematics in order to avoid the development of the atom bomb by German scientists.

Sincerely yours,
Werner Heisenberg

When I read this letter, especially the last sentence, I did not think it was "bizarre." I thought it was incredible. Heisenberg was always very careful not to make explicit claims like this. He let other people like Jungk and, posthumously, Powers and even his wife do it for him. Why would he make such a claim in such an apparently casual way? If he believed it, why hadn't he said so in any of his published accounts of his wartime activities? Then there is the phrase "falsified the mathematics." What does that mean? The atomic bomb does not involve some mathematical equation like $E=mc^2$ that you can "falsify." It involves hundreds and hundreds of engineering details. And there is also the mention of Hahn and Laue as co-conspirators. Hahn was, in the first place, not a mathematician but a physical chemist. While he disliked the Nazis, he liked his comforts even more. There is no indication that he ever risked his life—which is what this would have meant—for anything except helping Meitner escape Germany. He escorted her on a midnight train that crossed the Dutch border at a little-used point. Later she went to Sweden. But Meitner was so infuriated by his general laissez-faire attitude toward the Nazis that after the war she wrote him an exceedingly angry letter. And Laue, while he was at the Kaiser Wilhelm Institute, had nothing to do with the *Uranverein.* In any event, he was not a nuclear physicist. The whole notion of his involvement in this is absurd. Heisenberg would surely have known that anyone with the slightest knowledge of these matters who read this claim about Laue and Hahn would find it totally crazy.

Rose notes that there is something else that is odd about this letter. Mrs. Anshen gave her papers to Columbia University, including all her letters, but this one is missing. Rose remarks, "Nevertheless, it seems beyond doubt that the letter was genuine; its quotation in the Anshen memoirs is direct, and the letter itself is so startling that it could scarcely have been invented." I was so "startled" that I decided to try to check it out further. I have a colleague named Cathryn Carson who has been doing work on this history for several years. She has studied much of the Heisenberg *nachlass,* including the correspondence that is largely in the Heisenberg archive in Munich. I asked her if she had ever come across such a correspondence. She said that indeed she had, and that she would request permission of Helmut Rechenberg, who is in charge of the archive, to send me copies. Rechenberg and I have had strong disagreements about Heisenberg's wartime activities, but I am pleased to report that despite these disagreements he gave me permission to see the letters. He also gave Powers permission to see them since in his book he quotes from one of them. Whether Rose was given permission I do not know. There are in fact three letters.

The first letter is in English and is dated May 30, 1970. It is from Anshen to Heisenberg. In it she requests that Heisenberg write a review of Rabi's book, which appeared in her old series World Perspectives, for her new series Perspectives in Humanism. (In Heisenberg's alleged letter he does not even have the series right.) On June 19 Heisenberg replied in a letter of a page and a half in German. In this letter, addressed to "Frau Anshen" and not "Doctor Anshen," he asks her to inform Rabi that he is somewhat reluctant to write this review since he has strong disagreements with Rabi's Chapter 7, on the German atomic bomb project. He is especially upset that Rabi has accepted the arguments that were first put forward by Samuel Goudsmit in his book *Alsos* that the Germans would have been only too happy to have turned the bomb over to Hitler if they had had the competence to build it. Goudsmit had led a mission called Alsos—"grove" in Greek—to follow the advancing Allied armies into Germany and capture as many of the *Uranverein* as possible along with their equipment. In the event, nine members of the "club," including Heisenberg, Hahn, and von Weizsacker, were interned in England for six months in an estate near Cambridge named Farm Hall. Laue was also interned but mainly because it was thought that he might be better off in England than in postwar Germany. As it happened, British intelligence wired Farm Hall so that everything the Germans said to each other was recorded. In February 1992 the British finally released these transcripts; they are the most vivid record we have of the real thoughts of the *Uranverein*. Goudsmit had had access to them before writing his book. But Goudsmit got a certain number of things wrong, especially his statement that the *Uranverein* had not thought of plutonium. He and Heisenberg, and later von Weizsacker, got into a bitter exchange of letters and articles, so it is not difficult to see why Heisenberg objected to Rabi's reliance on Goudsmit's book.

The rest of the letter to Anshen is vintage Heisenberg. He constructs a sort of a murky parallel between the dilemma posed by handing an atomic bomb over to Hitler and that of using a hydrogen bomb to end the war in Vietnam, which was still going on. "Rabi," he writes, "completely overlooks the fact that the German physicists had about the same kind of psychological attitude towards putting a bomb in Hitler's hand as many Americans have today about the possibility of ending the American war with North Vietnam by dropping a hydrogen bomb on Hanoi." He ends the letter by asking Mrs. Anshen to take these matters up with Rabi so that they can avoid a public dispute. There is no mention of Oak Ridge and bomb damage in Germany. There is no mention of falsifying data, and there is no mention of Hahn and Laue. On July 9 Mrs. Anshen replied to Heisenberg, saying that she had spoken to Rabi and had decided that it might be better if Heisenberg did not review his book after all. She notes that ". . . Professor Rabi would not wish to enter into a polemical discussion with so great a physicist as you are. . . ." Knowing Rabi as I did, I can just see him concocting this phrase—"so great a physicist as you are"—with some glee. Rabi knew his man. He had offered Heisenberg a job at Columbia in 1939 when Heisenberg was on his last visit to America just before the war. Rabi told me that Heisenberg turned him down, explaining that he did not wish to lose his tenure in the German university system.

But what should we make of all of this? I know what I make of it. Unless and until someone shows me the original of the letter that Rose quotes, as far as I am concerned it is a chimera.

The second example of radical disagreement over the question of a German atomic bomb is more profound. I feel responsible for it. I was the one who first wrote about it, and I now think that what I first wrote was wrong, though both Powers and Rose continue to embrace it. I will explain. Beginning in November 1977, on and off for two years, I conducted a series of tape-recorded interviews with Hans Bethe which ultimately led to my three-part *New Yorker* profile. Bethe was head of the Theoretical Division at Los Alamos during the war and therefore in the inner circle of Oppenheimer's advisers. When we began discussing Los Alamos I asked Bethe if while they were working on the bomb they had had intelligence about what the German nuclear project was doing. Such intelligence, if it existed, was a very closely held secret during the war. I do not think, for example, that the people at Los Alamos knew about the activities of the Austrian-born chemist Paul Rosbaud. He remained in Germany throughout the war and passed information about the activities of the *Uranverein* through the Norwegian underground to the British. But Bethe did have some important information. He told me the following which I will give in the version he first gave to me and the version that indeed appeared in the *New Yorker.* This version I no longer believe to be entirely accurate.

Bethe reminded me that in September 1941 Heisenberg had come to Copenhagen where he met Bohr. The Germans had occupied Denmark in 1940, so that Heisenberg was a German visiting an occupied country. Indeed, one of the most distasteful of Heisenberg's wartime activities was visiting occupied countries. In December 1943, for example, he went to Krakow on the invitation of his brother's school classmate Hans Frank, who was then the governor general of Poland. One wonders what they talked about. Frank was then enthusiastically engaged in supervising the extermination of the Polish Jews. The ostensible reason for Heisenberg's visit to Copenhagen was as part of a German delegation, which included von Weizsacker, to a conference of astronomers organized by the so-called German Cultural Institute in Copenhagen, an outfit set up to disseminate German propaganda. Bohr boycotted this conference, and there was some question as to whether he would see Heisenberg at all. It should be remembered that while in Copenhagen in the late 1920s and early 1930s, under Bohr's relentless questioning, Heisenberg and Bohr hammered out what is known as the "Copenhagen interpretation" of quantum mechanics, which involves probabilities and uncertainties. This is what we still use and what we still teach our students. Bohr wanted at least to invite Heisenberg to dinner, but his wife Margrethe objected. She felt that the whole visit was "hostile." She never much liked Heisenberg. Still, Bohr managed to persuade her. After dinner he and Heisenberg had a private talk, the contents of which have become one of the most disputed aspects of Heisenberg's entire wartime activity. There is not even an agreement as to where it took place. Bethe told me that he thought it was in some "back alleys" in Copenhagen. Heisenberg has it as having taken place in either the Tivoli gardens or the Faelledpark. Others say it was in either Bohr's study at home or his office at his institute. Some have even claimed there was no dinner, that all the contacts between the two men were at Bohr's institute.

The content and purpose of the talk is even less sure. There seems to be no doubt that nuclear weapons came up. Bohr's son Aage, a Nobel Prize–winning physicist in his own right and his father's closest wartime confidant, has written, "In a private conversation with my father, Heisenberg brought up the question of the military applications of atomic energy. My father was very reticent and expressed his skepticism because of the great technical difficulties that had to be overcome, but he had the impression that Heisenberg thought that the new possibilities [perhaps an implicit reference to plutonium] could decide the outcome of the war if the war dragged on." But then what happened, or didn't happen, is one of the most mysterious aspects of the whole business. Bethe told me that Heisenberg gave Bohr a drawing of what purported to be the design of a German nuclear weapon. He said that later—I did not ask him how much later—the drawing was "transmitted to us in Los Alamos." I also did not ask what Bethe meant by "transmitted"—how? He told me that he and Edward Teller were asked to analyze the drawing and saw at once that it was a drawing of a nuclear reactor. Bethe said to me, "But our conclusion was, when seeing it, these Germans are totally crazy. Do they want to throw a reactor down on London?" My *New Yorker* article, which included this quote, was the first time anyone had mentioned such a drawing in print.

Both Rose and Powers have accepted this version of events but have drawn almost opposite conclusions from it. Powers is convinced that Heisenberg committed the act of a traitor—a traitor to Germany—in that he passed classified information to Bohr: "With this simple piece of paper Heisenberg had put his life in jeopardy." In other words, Heisenberg was not a Nazi collaborator but a Resistance hero. For Rose, showing Bohr this drawing was an act of intimidation, to convince Bohr that the atomic bomb would play a role in the forthcoming "Pax Nazica," Rose's term. In short, Bohr should now collaborate with the German scientists because he would soon have no other choice. Both Powers and Rose have persuaded themselves that Heisenberg really handed Bohr a secret sketch of a German nuclear weapon of some sort. But did he?

In 1994 I was paying one of my periodic visits to the Rockefeller University when Abraham Pais, Bohr's biographer, called me into his office. He had received a phone call from Powers some months earlier during which Powers asked him what he knew about the drawing. Powers had been taken aback when he had received a letter from Aage Bohr which stated flatly that "Heisenberg certainly drew no sketch of a reactor during his visit in 1941. The operation of a reactor was not discussed at all." Nor was plutonium, at least explicitly, since Bohr knew nothing about it until he was briefed about the Allied project in the fall of 1943 after he had escaped from Denmark. One would have thought that if Heisenberg had really wanted to communicate something important about the German program to Bohr, it would have been plutonium. After my talk with Pais *I* was taken aback. Indeed, I felt responsible for propagating a possibly serious error, and I decided to try to find out what happened. But how?

The first thing I did was to write to everyone I knew who had had a senior role at Los Alamos to see if anyone had heard of this drawing. Bethe of course had, and he repeated in writing what he had told me several years earlier. No one else seemed to know about it until I contacted the late Robert Serber. This turned out to be a gold mine. Serber was one of Oppenheimer's closest collaborators at Los Alamos. He also had a sense of history, so he held on to important documents. He also had an excellent memory. He told me what had happened, and he also sent me corroborating documents. When Bohr reached England in September 1943 he was briefed by the British on the progress of the Allied nuclear weapons program. Whether he told them what he knew about the German program we do not know. We do know that when he arrived in the United States in early December he met with General Groves and, it appears, showed him some kind of drawing. I do not know if there is anything in Groves's papers that documents this. But I do know that Groves was sufficiently alarmed by what Bohr told him that he alerted Oppenheimer.

On December 31, just after Bohr arrived at Los Alamos with his son, Oppenheimer called a meeting of a select group of staff members. Serber gave me a copy of the letter that Oppenheimer sent to Groves after the meeting which lists the attendees. I managed to contact all of those who were still alive. Aage Bohr was one, as were Serber, Teller, and Bethe. Serber recalled coming into the meeting a little late and Oppenheimer saying to him that they were discussing a proposal for a nuclear weapon that he said was Heisenberg's. He showed Serber the drawing, and Serber also realized that it was a reactor. No one I contacted could say whether it was claimed that this was a drawing made by Heisenberg or a drawing made by Bohr from memory. The drawing itself seems to have vanished. Bethe and Teller then wrote a report, of which Serber gave me a copy. They showed that this reactor could never explode like a nuclear weapon. No reactor can. This is what Frisch and Peierls had understood in 1940. A reactor operates with slow neutrons which have an enhanced probability of causing uranium or plutonium to fission. That is why the neutrons in a reactor must be slowed down or "moderated." The uranium in a conventional reactor is only slightly enriched with the U-235 isotope. A bomb, on the other hand, is made of nearly pure U-235 or plutonium, both of which are fissionable by fast neutrons. The entire explosive reaction in a bomb lasts only a hundredth of a microsecond.

Whether Heisenberg ever really understood this distinction is again one of the things that is vehemently debated. He said he did, and Powers agrees with him. Goudsmit and Rose say he didn't, and I agree with them. I base my position on an analysis of the Farm Hall transcripts. The Germans were recorded just after they first learned about Hiroshima. It is clear to me that they did not have the most rudimentary understanding of how a nuclear weapon worked. In a few days Heisenberg figured it out and gave his fellow detainees a lecture about it. The comments at this lecture make it clear to me that the Germans were hearing about this for the first time.

When I read the report that Bethe and Teller wrote, I realized that it was not any reactor they were analyzing but a particular design that Heisenberg clung to even though it had been shown by other more junior theorists in the *Uranverein* to be inefficient. Specifically this design involved layers of uranium metal submerged in "heavy water." This is water that consists of oxygen and heavy hydrogen, a rare isotope with one extra neutron. The Germans used heavy water to try to moderate the neutrons in their reactor designs. But it was very hard to come by, and the Germans never had enough. Fermi's reactor had lumps of uranium embedded in a purified graphite moderator. To me this meant that the drawing was something that came out of the *Uranverein.*

How can one reconcile this with Aage Bohr's absolute certainty—which he conveyed to me once again in a message delivered by Pais—that no reactor was discussed when his father met Heisenberg in 1941 and that no drawing changed hands. Here I will hazard a guess. I feel quite certain, unlike Rose and Powers, that Heisenberg never gave Bohr this drawing. It simply does not fit Heisenberg's character. He was never a Nazi, but he was a patriotic German. He told several people both during and immediately after the war that he had wanted the Germans to win it. He told Bethe that he wanted the Germans to win, because if the Allies won he was sure they would level Germany and destroy German culture. He was sure, he told Bethe, that once the Germans won the war the "good Germans" would take over and restore things to the way they were before the Nazis. What I think happened is that someone from the *Uranverein* visited Bohr and gave him this information. A likely candidate, it seems to me, is the *Uranverein* physicist Hans Jensen, who visited Copenhagen in 1942 and did discuss the German program with Bohr. Bohr was still persuaded that in any practical sense nuclear weapons were impossible, so he probably filed Jensen's report somewhere in his head. When he was briefed in England he recalled what he had been told. I cannot prove this, but neither can Powers or Rose prove their claim. It may be that some things about this history we will never know for sure.

When I first saw the Farm Hall transcripts in the spring of 1992 I thought they would make a great play if one could make the references and the physics clear. Unfortunately, from the historical point of view, the surviving transcripts cover only a minute part of the conversations that were actually recorded. The listeners, including General Groves, who was being sent them in Washington as soon as they were prepared, were interested in knowing what the Germans knew about nuclear weapons and whether they were likely to defect to the east. One of the copies I have has annotations by Groves. Every time the Soviet Union is mentioned there is a seismic reaction—underlining and exclamation points. He was seemingly not very interested in Hahn's almost suicidal reaction to learning that his discovery of fission had been used to destroy Hiroshima. Nor was he interested in the surreal award of the Nobel Prize to Hahn—and Hahn alone—for the discovery of fission in November 1945, while Hahn was still in detention. What the Swedish Royal Academy could have been thinking of I cannot imagine, and why Strassmann, Frisch, and Meitner were not included I cannot imagine either.

What a pity that we do not have a recording of what really went on between Heisenberg and Bohr in September 1941. One is free to let one's imagination wander. This is just what Michael Frayn has done in his play *Copenhagen*. Frayn is not a physicist, but he has evidently read a great deal about this. Some of it he had originally gotten wrong. Hahn did not fission uranium into two barium nuclei but rather into one barium and one krypton.

The Russians at Bohr's institute when Heisenberg was there were not Gamow and "Landers" but George Gamow and Lev Landau, who was arguably the most brilliant Russian physicist of his era and who later ran afoul of Stalin. These are minor and were corrected. What is not minor is a bit of dialogue that Frayn gave to Bohr. Referring to Heisenberg he says, "A White Jew. That's what the Nazis called him. He taught so-called Jewish physics. And refused to stop. He stuck with Einstein and relativity, in spite of the most terrible attacks."

It is true that Heisenberg, who like most Germans—indeed most of the *Uranverein*—was not a member of the Nazi party, was attacked by the Nazis for having done "Jewish physics." He was called a "White Jew"—what a strange epithet. But in July 1938, with the help of a family connection, he was vetted by Heinrich Himmler himself, who warned him to disassociate himself from the "personal and political attitude of the scientists involved." In other words, you could have relativity but not Einstein. This was codified in a conference in Seefeld in the Tyrolean Alps in November 1942 attended by thirty German scientists. Heisenberg and von Weizsäcker. Von Weizsäcker wrote the summary report which stated that "At the Seefeld meeting the opinion was expressed however that one must reject the imposition of the physical relativity theory into a world philosophy of relativism, as has been attempted by the Jewish propaganda press of the previous era." The report also said that "Einstein merely followed up the already existing ideas consistently and added the cornerstone." This does not sound like sticking to Einstein and relativity to me. Nonetheless I think Frayn has captured Heisenberg's moral ambiguity. He has resisted the temptation of having Heisenberg pass the drawing on to Bohr. His Heisenberg is neither a Resistance hero nor a simple Nazi collaborator; he is something more interesting and, indeed, more troubling. Frayn raises the question of why Heisenberg—and Bohr, for that matter—never did the relatively simple calculation that Frisch and Peierls did that showed that a bomb could be built. Certainly either of them could have done it. The sugges- tion of Frayn's play—and this may be part of the truth as well— is that somewhere deep in their psyches they were held back because they did not want to know the answer.

The Six Pieces of Richard Feynman

In the 1960s and 1970s, when I was writing long profiles of scientists for the *New Yorker,* it was frequently suggested to me that I do a profile of Richard Feynman. While Feynman was not then widely known outside the métier, he was universally known to all of us inside it. His Mozartean genius in physics seemed to be combined with an almost equally Mozartean urge to play the clown. We all knew that he had cracked safes at Los Alamos during the war and that he regularly performed in public on the bongo drums. Some of us knew of the tragic death of his first wife from tuberculosis and their communication games as they attempted to confound the security system at Los Alamos. His striking good looks and long hair made him seem like a somewhat superannuated flower child of the sixties—in short, a perfect *New Yorker* profile.

Nonetheless I did not try to write one. The reason is that I was concerned about Feynman's ego. All very good scientists have, in my experience, very large egos. This is not too hard to understand. Since they were very young they have been outstanding at doing something that is widely admired—the brightest in their class, brighter than their parents, brighter than anybody they know. It is bound to go to their heads. This kind of ego I could deal with. I understood it and respected it. But superimposed on this Feynman, it seemed to me from having observed him several times close up, had a second kind of ego, one I was not sure I could deal with. He had—with no disrespect intended—an actor's ego. He was constantly performing. When Freeman Dyson first met him in 1948 he wrote to his parents that Feynman was "half genius and half buffoon." In 1988, after Feynman's death, Dyson revised this to note that Feynman was "all genius and all buffoon." If you did not know who Feynman was and ran into him in a bar, say, you might well think he was either a stand-up comic or a New York taxi driver with an almost caricaturial accent. He did not have, or feigned not to have, any interest in "culture." That he would have read the *New Yorker* seemed inconceivable to me. That he would have had the patience to sit through the hours of interviewing—to say nothing of the endless fact checking—which doing a serious *New Yorker* profile then involved seemed equally inconceivable to me. I imagined myself trying to sit in rapt attention for days on end while Feynman "performed," then blowing it all by not being quite appreciative enough at some crucial moment and losing everything. I thought it was better not to try.

But around 1980 a colleague of mine who was then a junior faculty member at Caltech, where Feynman taught, and who saw Feynman on a regular basis, told me that Feynman was interested in having an article written about him and was, in particular, interested in having it written by me. This seemed quite incredible to me. While I had been introduced to Feynman several times, each time he had given me the impression of not having the foggiest idea of who I was. Still, as unlikely as my colleague's story appeared, obviously it was something that had to be followed up—so I called Feynman. It turned out that what my colleague reported was true, and Feynman explained to me the circumstances of his epiphany. He had been in a dentist's office, he said, where there were several copies of the *New Yorker*. Having nothing better to do, he started to browse through them. He assured me that otherwise he never read the magazine. As luck would have it, he had come across an article of mine in which I had referred to him as a "brilliant lecturer." God knows he was. I have never heard a better one in physics or anything else. His love of the subject and his love of performance produced marvels. You came away from a Feynman lecture feeling better about everything, no doubt about it. On the basis of this evaluation he had decided that I was, quite possibly, an acceptable Boswell.

That having been settled, it remained to establish a modus operandi. I was then teaching full time on the East Coast while Feynman was in Pasadena. I explained to him that in a few weeks I would have my spring break and could then come to California. That seemed agreeable, and he suggested that I stay at his house. While this was appealing in some sense, I did not think it was a very good idea. Unless you have actually conducted interviews like this, you will have no idea how much of a strain they are. You must constantly focus on getting a full and coherent story while at the same time attempting to make enough contact with your subject so that he or she thinks that what is happening is just a conversation. And this can go on for several hours at a time. I often come away from one of these sessions with a splitting headache. The last thing in the world I want to do is to carry on with the interview subject after hours. I need to be alone if for no other reason than to listen to the tapes I have made so as to decide how to proceed next. I made some explanation to Feynman having to do with not wanting to impose on his family life, and he said there would be no problem in locating me in a nearby motel.

So far so good. But then Feynman asked how much of his

time I would take in order to do these interviews. The only honest answer I could give him was that I did not know. I had done a profile of Lewis Thomas—not one of my better efforts—on the basis of one long luncheon interview (all the time he wanted to spare) and a few follow-up phone calls. But I had interviewed I. I. Rabi and Hans Bethe for well over a year before I had written anything. I had no idea where in this spectrum Feynman would fall. He seemed to accept this and told me that when the time came his secretary would arrange for the motel. I then proceeded to alert the *New Yorker* that a profile of Feynman might be possible and received a green light. But two weeks later Feynman called. He asked again how much of his time I would take, and I gave him the same answer as before. He then said, "But you told me it would only take a couple of hours." I replied that that was certainly not what I had said. He seemed to accept this, and as far as I knew our plans were still intact. Ten days or so before I was actually scheduled to go to California, I called Feynman's secretary to complete the arrangements for my motel room. "Oh," she said, "there is a film crew from the BBC coming at that time to film Professor Feynman, and he will not be able to see you." I gave up. Someone else could write Feynman's profile.

The first person to do so was Feynman himself, after a fashion. In 1985 he published *Surely You're Joking, Mr. Feynman! Adventures of a Curious Character.* These are "stories" he told Ralph Leighton, the son of a Caltech physics colleague, with whom he occasionally played music. Leighton taped his conversations with Feynman over a period of several years. This taping was surely going on before the time of my contact with Feynman. The book consists of what are essentially transcriptions of the tapes. The voice is authentic Feynman. It is all there. You can almost hear the New York accent. Whether the stories are factually accurate is another matter. I do know that one of the people whom Feynman discusses was so angry about Feynman's characterization of their joint work as being essentially entirely Feynman's doing that he threatened to sue. What a lawsuit that would have been! It is difficult to imagine a *New Yorker* fact-checker dealing with these stories—trying to locate, for example, the bartender in the Alibi Room in Buffalo, New York, where Feynman claims to have gotten into a fistfight with someone in the men's room. I doubt I could have brought forth these tales in a week's worth of interviews or even a year's. Likewise for those in the book's sequel, *"What Do You Care What Other People Think?"*, which was published shortly after Feynman's death. On the other hand, I doubt if anyone could get an idea of what made Feynman a great physicist on the basis of these stories. A great character perhaps, but a great physicist?

Since his death a number of books have appeared describing Feynman's life and work. (Of these my favorite by far is *QED and the Men Who Made It* by the physicist Silvan S. Schweber, published in 1994. Schweber has a deep understanding of the physics and the right sense of balance about the life. But, as perhaps is fitting, one must be a physicist to really understand the book.) In addition, a sort of minor cottage industry has developed which appears to consist of repackaging and recycling old Feynman lectures. (A similar industry has developed around Stephen Hawking.) One of the latest productions from the Feynman factory is called *Six Not-so-Easy Pieces.* If you strip away the jazzy title and the appreciative introduction by Roger Penrose, what you have precisely are six chapters that have been lifted bodily, without change, from Feynman's 1960s introductory course in physics, which was published in 1963 under the title *The Feynman Lectures on Physics.* These are lectures that Feynman gave at Caltech in 1961–1962 and then again in 1962–1963 to beginning students in an attempt to revitalize what had become, in some people's view, a dreary undergraduate program. If the aim was to inspire the undergraduates, they were not, as Feynman himself admitted, a great success. Attendance was high, but it consisted in considerable part of faculty members and graduate students who were happy to hear Feynman lecture on anything. The lectures are simply too sophisticated for even Caltech freshmen to follow. I cannot imagine anyone using Feynman's original book as a text in a course. I am sure, however, that many of us, myself included, have used it as a source when we came to preparing lectures on similar subjects for similar courses. His lectures are filled with nuggets of insight when you dig them out.

I had not read these lectures for several years, so when confronted with the recycled six chapters I decided that I would approach them as follows. I would pretend that they had come from an anonymous author who was trying to get a modern physics text published and who was presenting these six chapters, which do sort of hang together, as a sample. How would I react? The first thing that struck me is that they were evidently written by someone who was a master of the subject. This is something that a lay person might not be able to detect, just as a lay person might not realize how great a pianist was simply by the way he or she played the "easy" parts. In this kind of exposition you must know precisely what to leave out as well as include. I could tell by reading this material that whoever wrote it had a voluminous iceberg of understanding below the tip that was showing. Whether I could have guessed that it was written by Feynman is a different question. In 1697 Isaac Newton learned of a challenge problem that had been posed by the Swiss mathematician Johann Bernoulli. He solved it in a few hours and sent the solution—unsigned—to Bernoulli. Bernoulli knew immediately that it could only have come from Newton and made the famous Latin epigram "tanquam ex ungue leonem"—which, somewhat fancifully, might be translated as "the lion can be recognized by his claw prints." There is not a lot in these lectures that I could identify as Feynman's "claw prints." They certainly have the claw prints of a very good physicist, but not uniquely those of Feynman.

The second thing I noted about these lectures is that they are not self-contained. Terms like "Galilean transformations" are used without any prior definition or explanation. Worse, references are made—important references—to chapters of the original lectures which are not included in the six reproduced here. The reader is left suspended in midair. This might be understandable if what was being presented was really a sample of a promised whole, but considering that these six lectures are being packaged as an entity, it is quite inexcusable. It shows sloppy and indifferent editing and makes the whole project seem tacky.

The lectures also seem somewhat dated. The subject matter—the theory of relativity—is now part of "classical" physics, having been created at the beginning of the century. But things have happened in the forty-odd years since Feynman gave his lectures. Relativity experiments that Feynman says have not been performed have now been performed. There is even a remarkable set of experiments done by putting extremely accurate clocks on airplanes that show that one of the predictions Einstein made in his 1905 paper on relativity was wrong! He claimed that a clock on the equator would go slower than an identical clock on the North Pole—something that in relativity is known as "time dilation." He did not know at the time that there would be a compensating effect from the earth's gravitational field that would just cancel his prediction—something that has now been confirmed to high accuracy. Any good modern exposition of this subject would include these results. It would also include a discussion of cosmology. It was only in 1965—two years after Feynman gave his lectures—that Arno Penzias and Robert Wilson, then of AT&T Bell Laboratories, announced their discovery of what turned out to be the cosmic background radiation left over from the Big Bang. Before then cosmology, which depends fundamentally on Einstein's theory of gravitation that Feynman treats in these lectures, was something of a backwater in physics. Now it is of central importance. No modern textbook on this subject would fail to include a chapter on cosmology.

Finally, there is the matter of history, or what passes for history, in these lectures. To anyone who has studied the history of the invention of the theory of relativity, Feynman's comments on this subject, with all due respect, seem ludicrous. They simply show that he didn't care enough about the "culture" of physics to study its history with any seriousness. The least of the problems is that the only identification given to historical figures is, as rule, by the individual's last name and nothing else. Is the reader—in the first instance a college sophomore—expected to know that the "Poincaré" referred to only by his last name is the great French mathematician Henri Poincaré, who was a somewhat older contemporary of Einstein? Here again a conscientious editor could have filled in the reference without doing any disservice to Feynman's intent.

But there is worse. Feynman has got it into his head that Einstein, in creating his theory, followed a "suggestion originally made by Poincaré" concerning the principle of relativity—the notion that the laws of physics should appear the same to an observer at rest and an observer in uniform motion with respect to the resting observer. This proposition, which has its roots with Galileo, was emphasized by Poincaré in a lecture he gave in 1904 in St. Louis, Missouri. Apart from the fact that the twenty-five-year-old Einstein was then working in the patent office in Bern and had no access to this lecture, the relativity principle and its consequences were something that Einstein had been brooding about since he was sixteen. What he came to realize, which neither Poincaré nor anyone else had understood, was that it was impossible to maintain the relativity principle for both the classical mechanics of Newton and the electromagnetic theory of James Clerk Maxwell and others without some radical modification of our notions of space and time. What Poincaré and his contemporaries were trying to do was somehow to fix classical physics. Poincaré had an inkling that something radical might have to be done, but only Einstein, with no help at all from Poincaré, saw what had to be done and how to do it. It is a pity that no one challenged Feynman at the time and that this nonsensical view of the history was allowed to stand.

This brings me to the final question: who should buy and read this book? Alas, I think the answer is no one. The nonscientist will get very little out of these lectures. They were intended for very gifted Caltech undergraduates who already had some background in mathematics and physics. For a young person with such a background, why not get the complete set of lectures? They are difficult, but they can be grown into. Why settle for this poorly edited and incomplete sample? No reason I can think of.

Nash

Among mathematicians and theoretical physicists there can be a continuum of behavior that ranges between the profoundly eccentric and the truly mentally disturbed. To set the parameters, here are two cases: Isaac Newton and Wolfgang Pauli.

Pauli, who was born in Vienna in 1900 and died in Zurich in 1958, was one of the most distinguished of the quantum theorists. He was also incredibly eccentric—identical to his caricature, as Robert Oppenheimer once remarked. He had a massive head that bobbed rhythmically either up and down or back and forth, and he sometimes muttered to himself. I have no idea what someone who did not know who he was would have made of him, nor what they would have made of the scene I am about to describe.

Sometime in 1957, rumors reached us at the Institute for Advanced Study in Princeton, where I was enjoying my fellowship, that Pauli and Werner Heisenberg had found a Theory of Everything. If it had been Heisenberg alone, one might not have paid much attention, but Pauli was one of the most acerbic and brilliant critics of ideas in physics who has ever lived. Of some papers he used to say that they weren't even wrong. For Pauli to be smitten by such a theory meant that one had better look into it. In January 1958 Pauli came to the United States and gave a lecture on his revelation at Columbia University. Many of us came into the city from Princeton to hear him, including Oppenheimer and Niels Bohr, who was also visiting the Institute. After the lecture was over, Bohr was asked if he would like to comment. Before he could say anything, Pauli said that the theory might seem somewhat crazy. Bohr replied that in fact the trouble with the theory was that it did not seem crazy enough. By this time Bohr had gotten out of his seat in the front row and had begun to circulate around the long table in front of the blackboard on which Pauli had been writing. Pauli then followed him. When the two of them—titans of twentieth-century physics—reached the front of the table, Pauli again said that the theory was crazy and Bohr that it was not crazy enough. At the time it occurred to me that if someone had wandered in off the street and seen this, they might have had the two of them committed. Of course Pauli, to say nothing of Bohr, was not mad at all, just profoundly eccentric. Next let us turn to Newton.

I would imagine that the young Newton—the Newton of the "miracle year" 1666, when he made most of his great discoveries—might have seemed profoundly eccentric, at least very different. His amanuensis, one Humphrey Newton (no relation), some twenty years later said he had heard Newton laugh only once and that was when someone asked him of what use the study of Euclid might be. Newton's doctor, Richard Mead, when interrogated by Voltaire, disclosed that Newton had died a virgin, something that Voltaire then told everyone. But it does seem that Newton had one relationship that had the potential of becoming intimate. It was with a young Swiss savant named Nicolas Fatio de Dullier. Fatio became a Newton "groupie" and decided that he wanted to come to England to live with him. In the period between 1689 and 1693, at the times when Fatio was in London and Newton was in Cambridge, they exchanged letters of unusual intimacy for Newton. In one of them Newton even suggests that the two might share lodgings on one of Newton's forthcoming visits to London. But then sometime around 1693 Newton had a mental breakdown. He had paranoid delusions that his friends were plotting against him. He noted in a letter written in September 1693 to Samuel Pepys that he had "neither ate nor slept well this twelve month, nor have my former consistency of mind."

By the time this letter was written, Newton had recovered enough of his psychological equilibrium to acknowledge that he had been ill. Indeed, a few days after his letter to Pepys, he wrote one to the philosopher John Locke, which is surely one of the most remarkable documents in the entire Newtonian canon. Here is the pertinent part:

"Being of the opinion that you endeavoured to embroil me with woemen & by other means I was so much affected with it as that when one told me you were sickly & would not live I answered twere better if you were dead. I desire you to forgive me this uncharitableness. . . . I beg your pardon also for saying or thinking that there was a designe to sell me an office, or to embroile me with women. . . ."

Newton was in his early fifties at the time of this incident, and it is quite possible that even if it had not happened he would have made no further great contributions to science. But his scientific career did essentially stop at this point. So did his relationship with Fatio. They communicated sporadically but always with a remoteness on Newton's part. He had drawn back from something that he regarded as an abyss.

I do not know whether mental illness is more prevalent among mathematicians and theoretical physicists than among the general population. It is true that these scientists make their living by examining the contents of their heads. Perhaps, like ski instructors' knees, this activity exposes whatever weaknesses are there already. In any event, there are more recent examples. The nineteenth-century mathematician Georg Cantor, to whom we owe much of our understanding of mathematical infinities, from the age of forty on suffered depressions that were so deep that he often spent time in sanitoria. In 1918 he died in one. Earlier we studied the example of Kurt Gödel, one of whose great contributions to mathematics, ironically, concerned a problem—the so-called "continuum hypothesis"—on which Cantor had worked unsuccessfully. And now we have the case of John Nash, who is the subject of a recent biography, *A Beautiful Mind,* by Sylvia Nasar.

Before turning to Nasar's book, let me give the arc of Nash's life. I will fill in some additional details later. He was born on June 13, 1928, in Bluefield, West Virginia. His father, John Sr., had studied electrical engineering and worked his entire professional life for the Appalachian Power Company, so John Jr. grew up in an environment congenial to science. Curiously, for someone as gifted in mathematics as Nash seems to have been, he showed no apparent precocity. (Mathematicians often reveal themselves to be mathematicians as soon as they learn to speak.) At age thirteen or fourteen, however, Nash read Eric Temple Bell's book *Men of Mathematics.* It is a wonderful book for young people. I read it when I was a few years older than Nash was, but I have to admit that it did not generate in me the same ambition that it did in Nash, namely, to be the first person to prove Fermat's Last Theorem. I was satisfied to understand what the theorem was.

In his senior year in high school Nash entered the George Westinghouse science competition and won a scholarship to college. Nash chose to use his scholarship to attend the Carnegie Institute of Technology in Pittsburgh, where he intended to become a chemical engineer. Curiously, John von Neumann, the great Hungarian-born mathematician whose work was to prove so important for Nash, graduated in 1926 from the Technische Hochschule in Zurich, Einstein's old alma mater, with a degree in chemical engineering. Of course the *same* year von Neumann also received his Ph.D. in mathematics from the University of Budapest, having produced a dissertation on the foundations of mathematics, the one that was outshone a few years later by Gödel.

Nash made a sufficient impression on his professors at Carnegie so that he had a choice among the best graduate schools. By this time he had decided to become a mathematician. He chose Princeton because it offered a somewhat larger fellowship than Harvard did. Not only did Princeton have a superb mathematics department of its own, but there was the accessibility of the faculty at the Institute for Advanced Study which, at the time, included both von Neumann and Gödel. Nash arrived in Princeton in 1948, four years after von Neumann and the economist Oskar Morgenstern had published their seminal book *The Theory of Games and Economic Behavior.* Game theory was probably a more active subject at Princeton than anywhere else at the time, and Nash became interested in it. Indeed, he wrote his Ph.D. thesis—all twenty-seven pages of it—in 1950, on game theory. This is one of the things I will amplify later.

Nash had hoped to be able to remain at Princeton, but he was not offered a job, so he took an instructorship at MIT, which was in the process of building its mathematics department. In June 1951 he moved to Cambridge. By this time he had already spent the summer of 1950 in Santa Monica as a consultant at the RAND Corporation, the defense think tank that Stanley Kubrick referred to as the BLAND Corporation in *Dr. Strangelove.* It was in the summer of 1952, while consulting at RAND, that Nash probably had his first homosexual relationship. But that fall, on his return to MIT, Nash began an affair with a somewhat older woman named Eleanor Stier, with whom he had a son, John David Stier, in 1953. Nash had no interest in marrying Eleanor and little interest in his son. He seems also to have continued to have homosexual relationships and, indeed, in the summer of 1954 he was arrested for indecent exposure—a misdemeanor—in a men's room in Santa Monica. This cost him his security clearance—he was consulting at RAND—and he never again worked for the government, though MIT, to its credit, did not threaten to take away his job, and there were no other adverse consequences. One may contrast this with what happened to the British mathematical genius Alan Turing, who was arrested in Great Britain in 1952 for a violation of the Gross Indecency law of 1885—the same one used to convict Oscar Wilde—and forced to take estrogen, which at the time was thought to "cure" homosexuality. Turing, as we have noted, committed suicide in June 1954 by eating a poisoned apple.

Perhaps a competent psychiatrist who had had the opportunity to study Nash's behavior at this time—which even excepting his sex life, was becoming decidedly more peculiar—might have realized that something was seriously wrong. Certainly his colleagues thought he was becoming increasingly eccentric. But MIT already had one of the great eccentrics of all time, Norbert Wiener, on its faculty, and in comparison Nash must have seemed relatively normal. Certainly Alicia Larde, who was one of the few coeds at MIT in the 1950s, thought so. She fell in love with Nash, and after an erratic courtship which included a very emotional encounter with Nash's former mistress, Eleanor Stier, they were married in early 1957. What exactly she knew about Nash's ambivalent sexuality—he had another homosexual relationship in the summer of 1956—is not clear. But it is clear that by 1959 Nash had crossed over the boundary from deep eccentricity to serious mental illness. For the next thirty years he was in and out of mental institutions and for all practical purposes had disappeared. Then, beginning in the late 1980s, Nash, who had been living in Princeton but was without a job, had a seemingly spontaneous remission from what had been diagnosed as paranoid schizophrenia, something that happens in less than 10 percent of such cases—perhaps Newton's was one of them.

In the fall of 1989 a young Swedish economist, Jorgen

Weibull, paid a visit to the Princeton campus where Nash had been leading a kind of ghostly existence. Nash had a habit of leaving strange, though nonthreatening, messages on peoples' blackboards. Very few people at Princeton knew who he was. Weibull did, and the chairman of the mathematics department persuaded Nash to have lunch with him. I am not sure if the chairman knew the reason for Weibull's interest. Nash certainly didn't. Weibull was representing the Royal Swedish Academy of Sciences, and he had been charged with the mission of finding out whether Nash's mental health was sufficiently sturdy to stand up to the rigors of being awarded a Nobel Prize in economics—basically for his twenty-seven-page Ph.D. thesis, which had by now become an essential element of mathematical economic analysis. Weibull decided that Nash was not substantially stranger than the other Princeton mathematicians he encountered. In October 1994 it was announced that Nash was to share the Nobel Prize for Economics with Reinhard Selten and John Harsanyi. The morning after the announcement, a colleague of mine at Princeton, who knew who Nash was but had never really talked to him, saw him walking on the campus. He went up to Nash to congratulate him. Before he could say anything, Nash said, in total sincerity, "You are famous. I saw you on television." My colleague had appeared briefly on some television news program to comment on a new discovery in cosmology. That is the arc of John Nash's life. I now turn to Nasar's book.

In reading it I found myself almost haunted. I am two years younger than Nash, but we seem to have traced many of the same routes. In 1951, when Nash arrived at MIT, I was finishing my senior year at Harvard and was a mathematics major. Many of the people Nasar mentions were my teachers and colleagues. Marvin Minsky, one of the founders of artificial intelligence, had just come back to Harvard from Princeton where he had been a contemporary of Nash. Minsky, who is mentioned frequently in Nasar's book, was one of the people who never deserted Nash even in his darkest times. Minsky recently told me that while he was at Princeton he was stuck on a theorem he needed for his thesis. Nash gave him a suggestion that was remarkable because it seemed to have nothing to do with his theorem but turned out to be the key to proving it. He also told me that during this time he and several other mathematics graduate students, including Nash, from time to time used to eat out in one of the restaurants in Princeton. If the service was really bad, Nash would wait until the tip money was on the table and then pocket some of it, something he called a "negative tip." Minsky said they would then complain, "But that's our money!" "No it isn't," Nash would reply, "It's her money now." It posed an interesting intellectual dilemma. Forty years later, when Nash won the Nobel Prize for Economics, Minsky decided that he must have been right.

George Mackey, who visited Nash after he had been committed to McLean Hospital in the spring of 1959, was my first great mathematics teacher. Nasar writes that Mackey asked Nash how he could believe that extra-terrestrials were sending him messages, to which Nash replied, "Because the ideas I had about supernatural beings came to me the same way that my mathematical ideas did. So I took them seriously." Richard Palais, who is a professor of mathematics at Brandeis, and who found Nash a temporary job there in 1965, was one of my roommates when I was a graduate student. Nasar describes people that I have not seen for forty years, bringing back a flood of memories.

Then there are the places. The last time Nash consulted at RAND was the summer of 1954. I consulted there four years later, and the people and the atmosphere were unchanged, all the way to the presence of Admiral Chester W. Nimitz's daughter Nancy, with whom I used to play tennis. Nash spent the 1956–1957 at the Institute for Advanced Study. I arrived the next year. Nash had a run-in with Robert Oppenheimer about the quantum theory. Nash seems to have spent his time at the Institute looking into the foundations of the quantum theory, usually a disastrous enterprise for mathematicians. Von Neumann was one of the few who did it with any success. George Mackey is another. About the time of Nash's visit, the physicist David Bohm, who was then at Princeton, had published some papers on "hidden variables" in quantum theory. These are supposed to be variables that underlie the theory and make it appear to behave more classically. From Nasar's book, my impression is that this is what had caught Nash's fancy. He tried to talk to Oppenheimer about it and was given the "treatment." Oppenheimer could be very nasty when he thought his time was being wasted. For Oppenheimer, Niels Bohr had solved all the problems in the foundations of the quantum theory, and Bohm—and certainly not Nash—had nothing to teach him. But it was also characteristic of Oppenheimer that in 1963 he helped Nash with a one-year appointment at the Institute, using the Institute's own funds, with the hope that being there might help Nash to restore his psychological equilibrium. I even thought I might have taken a course from Nash at MIT when I was a graduate student at Harvard, something that we often did. Nasar gives a list of the courses that Nash taught, and they do not correspond to what I remember having taken. Also, her pictures of the young Nash do not ring a bell. He was strikingly handsome and very athletically built. The older Nash does look a little more familiar, and I wonder if we ever met during my many visits to Princeton.

It is to Nasar's great credit that she was able to interview so many of the mathematicians who knew Nash at various times in his life. It is a pity she did not get one of them to look over her manuscript carefully for mistakes. Nasar is an economics correspondent for the *New York Times,* so she is qualified to write about that side of Nash. Mathematics is something else altogether. Let me give two examples—not very important by themselves but symptomatic of what I think is a serious defect in her book. In describing the ambiance at MIT when Nash was first getting to know his future wife Alicia, Nasar quotes a woman called Emma Duchane, who was a physics major at the time. Duchane says, "We wanted intellectual thrills. When my boyfriend told me e equals I to the PI minus 1, I was thrilled. I felt the absolute joy of the idea." Insofar as I can make any sense of this statement, it is false. Perhaps Duchane's boyfriend was having her on or perhaps he told her the correct formula "e to the (pi × i) = −1," $e^{i\pi} = -1$. Equally false is her statement a few pages later that Einstein employed the non-Euclidean geometry of the nineteenth-century German mathematician Bernhard Riemann "in formulating his special theory of relativity." The special theory of relativity, which Einstein created in 1905, had nothing to do with Riemann's geometry. It takes place in a flat Euclidean world. It was only the general theory of relativity and gravitation, which Einstein published in 1916, in which non-Euclidean geometry became relevant.

These and other mathematical howlers are not by themselves all that important, but they show an unfamiliarity with this intellectual world, and that is significant for Nasar's book. Mathematicians are certainly very intelligent, especially if intelligence is defined as that which mathematicians do well—abstract thought. If one is not used to them, it is easy to get the impression that every mathematician is a "genius." The whole notion of genius becomes dumbed down. To Nasar, everyone she writes about seems to be a genius. Perhaps she should introduce a genius hierarchy, something like the one that Cantor invented to classify infinities.

It is particularly difficult to classify Nash. For all intents and purposes, his career in mathematics stopped when he was thirty. The last thing he was working on—a conjecture in number theory by Riemann—to this day has not been proved. From Nasar's account it would appear that Nash's breakdown was beginning to manifest itself in his approach to this problem. There was almost the desperation of someone who felt that his career was about to end and that he had to do something spectacular before it did. His colleagues tried to tell him that the approach he was using had already been tried and didn't work. Nash did not listen; not long after, he broke down. Certainly the work he did during the brief time that he was active in research—during his whole career he published only some twenty papers—was truly first-rate. But how do you compare it with what a Gödel or a von Neumann managed to do, even before they were thirty? In her attempt, understandable perhaps, to elevate Nash, Nasar occasionally borders on the silly. She describes a lecture that Nash gave, his first in thirty years, after he won the Nobel Prize, at the University of Uppsala on cosmology, something he had been interested in before he became ill and had taken up again. (A colleague of mine at Princeton tells me that Nash comes to cosmology colloquia there and asks very sensible questions.) She writes that Nash's talk "began with tensor calculus and general relativity—stuff so difficult that Einstein used to say he understood it only in moments of exceptional mental clarity." With all respect due, the hundreds of people, from graduate students on upward, who now work in cosmology use tensor calculus and general relativity on a daily basis. You can't do without it. While it was certainly exotic when Einstein introduced it in 1916, in the years since it has simply become a standard part of the curriculum.

In preparing this essay I did something that I thought I would never do in this life. I bought a textbook in economics. It is called *Game Theory with Economic Applications* by H. Scott Bierman and Luis Fernandez. It was published in 1993, the year before Nash won his Nobel Prize. I bought it because I wanted to see if what Nash had done in his thesis had in fact gotten into the mainstream of economic theory—became a textbook subject—and because I wanted a somewhat more technical and more detached account than the one that Nasar offers in her book. I report on my findings with a good deal of reticence and with many apologies to the experts. I am still studying my textbook. One thing I can tell you for sure is that Nash's work has become a textbook subject. In fact my textbook has two long chapters called "Nash Equilibrium I" and "Nash Equilibrium II," which are pivotal. I also noted that the 1965 edition of the *Encyclopaedia Britannica* has an excellent entry on game theory—there is an even better one in the 1987 edition—which has a paragraph on Nash equilibrium. I was quite surprised to see this entry in the 1965 edition. It makes it clear that while Nash himself had vanished by this time, his work hadn't. Let me now attempt a brief explanation of what Nash did.

The analysis of strategies for playing games—at least games whose rules can be formalized mathematically—goes back a long way, but the modern theory starts with the work of von Neumann, which he first began in the late 1920s when he was about the same age that Nash was when he did his. Von Neumann's analysis dealt basically with two-person, zero sum, noncooperative games, subject to clearly specified rules. By zero sum one means that one player's loss is necessarily the other's gain, and by noncooperative one means that the players do not enter into binding agreements with each other before the game starts. The strategy of a noncooperative game evolves as a function of what happens while the game is being played. Von Neumann proved what is known as a "minimax theorem." The essence is this. Suppose there are two players, say Mary and Tom—my textbook likes Mary and Tom—and they are playing such a game. Mary wants to choose a strategy that will maximize her gain, taking into account that Tom is smart enough to see through this strategy and will counter with one of his own designed to minimize Mary's gain. Thus Mary will look for a strategy that minimizes the damage that Tom can do, in other words a strategy that will maximize Tom's chances of doing minimum damage—a "maxmin" strategy in the lingo of the subject. Conversely Tom will look for a strategy that minimizes the losses that Mary can inflict. What von Neumann showed was that for this class of games such a strategy must exist. Finding it in a given game is another matter, and the theorem won't help you there. It was this theorem that became the basis of what von Neumann and Morgenstern did in their 1944 book. Indeed, not much progress in game theory had been made from the time of von Neumann's first papers and 1950 when Nash entered the picture.

What Nash did goes beyond simply allowing one to set aside

the assumptions of von Neumann's theorem, such as that the game has to be zero sum. It also changes the paradigm. To see how this works I will consider, at least in part, an example stolen from my textbook. In a community there are two automobile dealers—the textbook calls them Honest Ava and Trusty Rusty—each selling the same kind of automobile. Between them, then, they have a monopoly. But they are not allowed to cooperate. They want nonetheless to set prices. In this "game" they have three possible prices to chose from—high, medium, and low. How should they set these prices to produce a situation that both maximizes what they can make by selling the cars and is "stable," in the sense that having set them this way they don't immediately have to reset them in response to the competitor's defensive action that forces them to do so? If there is such a strategy, it is called a Nash equilibrium. Crucial in this discussion, as I am going to illustrate, is that the competitor acts in a rational way to maximize his or her benefit. If a Nash equilibrium is reached, everyone will have done the best that can be done. As my friend Yvon Chouinard, who founded Patagonia, once said about rock climbing, "There's no winners and there's no losers, but you can't get out of the game."

If Ava and Rusty could collude, the strategy would be obvious. Both of them would set their prices as high as possible, and the story would be over. But they can't, so here is how Ava will reason. "If I set my price high and Rusty acts rationally, he will lower his price to medium. But he will reason that I will then act rationally, which means that I will respond by setting my price low. Hence setting my price high cannot lead to an equilibrium."

One can now go through the rest of the cases and show that the only equilibrium solution is for both Ava and Rusty to set their prices low from the beginning—a wonderful arrangement for the consumer. This was Nash's basic insight, the notion of equilibrium. What he showed was that in a very wide range of such "games" there must be at least one such equilibrium strategy. There may be several, in which case one must decide among them as to the best. The theorem is nonconstructive in the same sense as von Neumann's was. It doesn't tell you *how* to find the equilibrium strategy, just that one exists. Perhaps that is why, according to Nasar, von Neumann was dismissive when Nash went to see him after having found his result. More likely, by 1950, six years after his work with Morganstern, von Neumann had lost interest in the subject. By 1950 von Neumann was deeply involved in creating the first modern electronic computer—the EDVAC—and game theory may have seemed a little trivial in comparison. But now we have the computer and Nash's equilibrium theory, both used routinely by economists. The adumbrations of Nash's thesis have become a cottage industry.

To me, one of the most remarkable parts of Nasar's book is her revelation of the machinations that went into Nash's Nobel Prize. Having tried to do something like this with the curious 1944 solo award to Otto Hahn for the discovery of fission—the work was done in collaboration with Fritz Strassmann and correctly interpreted by Lise Meitner and Otto Frisch, none of whom were recognized—I have firsthand experience of the difficulties one runs into. There is a fifty-year rule, which in the case of Hahn expired in 1994, before I tried to learn what happened. Nonetheless I got nowhere. Furthermore I was told by people who had had access to some of the deliberations for earlier prizes that they were in Swedish and simply summarized, without details of who said what. But Nasar has the whole messy story.

In the first place, the Nobel Prize for Economics is financed by the Central Bank of Sweden and not by the Nobel Foundation. Thus the money does not come out of Alfred Nobel's will. Indeed, its correct name is "The Central Bank of Sweden Prize in Economic Science in Memory of Alfred Nobel." (There is no Nobel Prize for mathematics.) But the prize is administered by the Royal Swedish Academy of Sciences and the Nobel Foundation. By 1994, when the prize committee had decided that a prize in game theory was appropriate on the fiftieth anniversary of von Neumann and Morganstern's book, a good deal of opposition had developed as to whether there should be a "Nobel Prize" for economics at all. Some scientists felt that it was a bad precedent that could dilute the significance of their prize. This became part of the politics of the 1994 prize. Then there was the matter of Nash himself. Not only had he done nothing in the field for forty years, but one of the committee members argued that he had been a consultant at the RAND corporation and that they had worked on bombs—a criterion that would have ruled out a great many physicists, from Feynman on down. Nash's history of mental instability brought sympathy but also concern. Various other candidates were suggested, but in the end, by a narrow vote, Nash was named. What it meant to Nash can only be imagined.

This brings me to the last thing I want to discuss, Nash's recovery. I had the chance to have a brief conversation with Oliver Sacks about this. He was familiar with Nash's case history. He thought it was crucial that during the period of his recovery Nash had a supportive environment. It is also fortunate, I think, that Nash was not subjected during his hospital stays to something like electric shock treatments. The treatments he did undergo were bad enough. Nash and his wife Alicia were divorced in 1962. He spent much of the next two years in the Carrier Clinic in New Jersey. After some years of desultory wandering he returned to Princeton in 1970. By this time he had nowhere to go, and, motivated by pity, love, and perhaps hope, his ex-wife let him move back in with her. He still lives with her, though they have never remarried. In 1959, when Nash was in the McLean clinic in Boston, Alicia gave birth to their son, John Charles Nash. He was also part of the household that Nash returned to in 1970. This was when Nash began haunting the Princeton campus. Here again he was fortunate. No one harassed him. He was even allowed to have a free account on the computer. At first he seemed to be using the computer to continue the rather crazy numerology he had been engaged in—"making names out of numbers and being worried by what he found," as Hal Trotter, who saw

Nash nearly every day at the computer center, recalled. "Gradually that went away," Trotter noted. "Then it was more mathematical numerology. Playing with formulas and factoring. It wasn't coherent math research, but it had lost its bizarre quality. Later it was real research." Nash was getting well.

No one can say if this remission is permanent. One can only hope so. What is almost unbearably sad is that his son John Charles, brilliant in his own right, has inherited the same sort of mental illness that his father had, and has had it for a larger fraction of his life. Most of Nash's energies are now taken up trying to help his crippled son.

Index

A NOTE ON THE AUTHOR

Jeremy Bernstein was born in Rochester, New York, and educated at Harvard University. He is a theoretical physicist who is now professor emeritus of physics at the Stevens Institute of Technology. He has held appointments at the Institute for Advanced Study, the Brookhaven National Laboratory, CERN (the European Organization for Nuclear Research), Oxford University, the University of Islamabad, and the École Polytechnique. He has written some fifty technical papers and three technical monographs, most recently *Kinetic Theory in the Expanding Universe* and *An Introduction to Cosmology.*

Mr. Bernstein is perhaps best known for his writings for non-scientists, including *Albert Einstein* (nominated for a National Book Award); *Dawning of the Raj; Three Degrees Above Zero; Mountain Passages; Science Observed; The Tenth Dimension; Cosmological Constants; Quantum Profiles; Cranks, Quarks, and the Cosmos;* and *Hitler's Uranium Club.* From 1961 to 1993 he was a staff writer for the *New Yorker.* Among a great many honors and awards, he has won writing prizes from the American Association for the Advancement of Science, the American Institute of Physics, and the American Alpine Club. He lives in New York City and Aspen, Colorado.